The Lost Cowboy

J.B. Zielke

The Lost Cowboy
© and published 2022 by J.B. Zielke
All Rights Reserved

ISBN: 979-8-218-11339-1

The events in this book are the real experiences of J.B. Zielke, however some names and places have been changed to protect people's privacy.

Contact the publisher at: zielkelivestock@gmail.com

Cover art by Taylor Jackson

Interior and cover design by Alane Pearce, Pearce Writing Services LLC contact at apearcewriting@gmail.com.

Acknowledgements

This book would not have been possible without the many hours of work from my dear friend Sarah. From simple editing to computer wizardry, Sarah had a hand in every step of this book. I must also thank my friends and family who helped me through all the trying times overseas. They watched my dogs, allowed me to sleep on their couches or front lawns and gave me work when I was too broke to buy my next meal. Lastly, I must thank all the kind souls I met abroad who took a chance on a stranger. It takes a special type of person to invite a dirty traveler into their business and family.

Table of Contents

Sweden

South Africa

Mexico

Mongolia

Foreword

I first met J.B. Zielke in late January outside the Star Hotel in Elko, Nevada. We were both there for the National Cowboy Poetry Gathering. He was there for the music, and I was about to play a set. He told me he was fresh off finishing his goal of cowboying on every continent with cattle. We found ourselves swapping stories back and forth. He told me about riding reindeer in Mongolia, wild African horses, and hiding from stray bullets in Mexico.

In a world where the wild places are shrinking, J.B. has traveled to learn, swap knowledge, and brave all the world's methods of working cattle and horses. From catching wild bulls on motorcycles in Australia to being chased by monkeys while on horseback in Argentina, his stories are filled with heart-pounding close calls and mistakes. Through lessons learned the hard way, he writes about the challenges cowboys and ranchers face all over the world. In his book, J.B. dives into some of the last traditional cowboy cultures and how they are surviving and adapting in the face of modern machinery and mass production.

As cattlemen, we dream of what life was like on a trail crew driving herds north with no fences in sight. Places where technology fades away, and the spirit of the wild is still king. Something has driven J.B. to travel to where the modern world has just begun to meet the untamed land. In this book, he records what those worlds are like while they still exist.

Colter Wall

Australia

1
The Beginning

The nerves twisted up my gut like a rattlesnake, unlike the nerves of a job interview or meeting your girlfriend's parents. There was something far edgier about these nerves. It was the feeling I used to get before climbing on the back of a bucking bull. The feeling of kissing a pretty girl for the first time. As if something as small as the wind changing directions could cause a catastrophe. It churned deep inside me, a feeling that I would never shake.

I had arrived in the little Aboriginal community of Kowanyama. The name means "Land of Many Waters." I had flown two hours from Cairns deep into the bush on a six-seater bush plane. The other two men on the aircraft giggled and gave me a funny look when I told them where I was going and what I was doing. It was the same look I had received from my friends when I was about to do something stupid. Sure, they did not want to see me die, but they would laugh when I failed. These men were both contractors, building housing developments in rural communities like the one I would live in. The little bush plane was like a school bus. It flew in a large circular route, letting people on and off at each small community it landed in.

When I jumped out of the plane onto the dirt runway, one pilot followed me out. He opened the compartment below and grabbed my backpack and suitcase. He chucked them on the ground and said nothing more than, "Good luck, mate." Before I could step off the airstrip, the plane was at full throttle. I caught a few pebbles and plenty of dust in the face. The only building in sight was the one I dragged my worldly belongings toward. I had seen an old Aboriginal man driving a tractor to refuel planes. He had ducked into the building when I landed. When I approached and asked to use the landline in the building, he looked up at me, shocked. It was rare to see a white man in this part of Australia. It was unheard of for that white man to have an American accent.

Before I left the United States, they gave me a couple of numbers to call when I landed. The old man dialed them and got no answer on either of them. My heart sank. I had flown to the other side of the earth, to one of the most dangerous wild parts of Australia. Now I was standing in the intense burning sun, unsure if I was even in the right place.

My soon-to-be boss, Lachie, was the brother of a friend I had met in college. My contact with him had been extremely poor before leaving home. I would send him a long email asking many questions. He would respond with one or two sentences total because he was using a satellite phone to email from. He gave me a time and place to be and said little more than that. Now I was at the location, and he was nowhere in sight.

The old man at the airport asked who I was working for. When I told him everything I knew about the owner and the company, he said, "Never heard of him." He grinned, revealing his three or four remaining teeth and disappeared into his building, leaving me sitting on my bags in the sun, terrified. After an hour, I heard the rumble of a diesel engine. A silver Toyota pickup truck, referred to as a "ute," pulled into the parking lot. A cloud of dust covered me and everything around me. A young man about my age who introduced himself as Archie walked right at me, shouting out, "How ya going, mate? Are you the American?" I had never been so relieved.

Another man stepped out of the ute, looking slightly older than me. He introduced himself as Tommy, the contractor that worked for the man I had been emailing. They were both hard men, almost like something out of a movie. Living and ranching in the Rocky Mountains in the United States, I had been around my share of rough and wild cowboys and rig hands. These guys were different. Neither of them wore shoes. They wore button-up shirts with no sleeves very short dirty shorts, and cowboy hats that looked like a truck had run them over. You could tell that they both had been working in extremely dusty conditions. They wore a fine layer of dust over every inch of their body.

After smashing my bags in the back between the leaky fuel drums, tools, car parts, and cattle working equipment, we drove off to town. We made a quick stop at the gas station, which Australians call a servo. It was about nine in the morning; a drunk Aboriginal man stumbled up and started talking with Tommy and Archie. They explained to him that I was American.

He walked up to shake my hand and slurred into my face, "An American, in far north Queensland? How deadly is that, bro?" This was an expression I had never heard before. I thought in some roundabout way he was telling me I would die in this place. I later discovered the word deadly was interchangeable with awesome. How right he was. For the next three months, I was doing something deadly every day.

The tiny town was the only civilization that existed out there. We hopped back into the ute after we had filled up with diesel and filled all the fuel cans. Then, we started out into the bush to our camp. The camp was almost an hour from the tiny Aboriginal community I had flown into. I noticed four-wheelers, dirt bikes, and vehicles scattered everywhere upon arrival. We even had a massive six-wheel-drive army truck with a custom cage on the back. It could haul horses or bulls in or out of anywhere.

My favorite vehicles in the camp were the bull catchers: old Toyota SUVs and pickups that were modified into wild animal catching machines. Bull catchers had thick steel plating protecting the sides, and massive steel bumpers on the front with tires strapped on as padding. I thought of the apocalyptic movies, the end of the world. People fabricating giant war machines built to withstand anything. These were the cowboys of the apocalypse.

The camp was centered around the portable kitchen. They had placed the kitchen in a half-collapsed tin building. Old hay filled the shed, a paradise for the hundreds of types of deadly snakes in Australia, snakes we would encounter often. Most of the other crew members slept in the back of their utes. Others slept on a cot to keep themselves off the ground.

We had a shower that was a hose and a couple of walls to protect your privacy. I was informed immediately that you must always examine the shower before use. Poisonous serpents enjoy the cool, damp environment it provides. They gave me a "swag," or some might call it a bedroll or sleeping bag. They told me to throw my swag down where I liked. Unfortunately, I had no cot or ute to keep it off the ground until I found something.

After being introduced to the people in the camp, I was told the rest of the crew was already out catching bulls. I needed to get work clothes on and head out with the rest of the crew. I would jump in with the bull truck driver. He was a middle-aged man with tattoos from his ears down to his fingertips. He was missing his front teeth, and everyone called him "Mad Rich." I followed my orders, and a girl named Kelly jumped on with me.

Kelly was the only girl in camp. It was rare to see women in this line of work.

Before leaving camp, I was told to climb onto the roof of the truck, as it was cooler and quieter. It was a cab-over truck, meaning the engine sat between the driver and passenger seat. The engine was completely exposed. A body part too close to it could get it sucked in or ripped off. It was sweltering and deafening to sit in the cab of that truck. Mad Rich did it with a broken-toothed smile while he smoked his hand-rolled cigarettes.

The truck screamed at top speed down rough country roads made from bulldust: dust so fine that it could take 20 minutes to settle after a vehicle drives past. Kelly and I held on for dear life on the roof of the massive, thundering beast of a truck. It would take only one quick stab at the brake pedal or a big bump to launch Kelly and me off the top of the truck. The speed we were driving at, coupled with the massive height of the truck, would have killed us if we fell.

There was something profoundly rewarding about that moment. It was the first of countless times I would be inches from death but somehow find the bliss in it all. Half an hour into our bone-jarring ride, my mind had drifted off somewhere else. The heat, the dust, the speed, the danger, and the screaming of the old worn-out engine faded.

For the first time in a long time, I felt I was free. Hunter S. Thompson said, "Faster, Faster, until the thrill of speed overcomes the fear of death." This is how I wanted to live my life. I was in the middle of an adrenaline-fueled camp of modern-day outlaws. I would leave Australia with an entirely new level of addiction to adrenaline.

Just then, Kelly grabbed me. She pulled me to the side as a tree branch the size of my arm smashed into the front of the truck, nearly ripping me off the top. Then, I snapped back to reality; we had arrived in a clearing. Thousands of cattle roamed around in dead, dry grass as far as the eye could see. We pulled up to a group of three or four men standing around a wild bull. They tied his huge, sharp horns to a tree with a thick cotton rope. I was about to learn some things they do not teach you in school.

2
Catching Feral Cattle

When we pulled up to the big, red bull tied to the tree, it was clear he had spent his entire life living on the arid, treeless flats of Kowanyama. He was healthy but thin. I could tell a person had never laid a hand on this wild bovine before. I could tell by his demeanor and unmarked skin. His horns were long, sharp, and deadly. His hide was shiny but covered in dust and flies. He was free from any brand or marking.

An animal lacking an owner's mark has long been a symbol of a renegade, an outlaw, a symbol of the last wild things. In Australia, they referred to cattle lacking a brand as "clean skins." In the United States, you will hear them called mavericks. They are fascinating animals. In appearance, they look like normal domestic cattle. However, the demeanor of the wild cattle fascinated me. This animal had never felt the touch of a human. The closest it had ever come to humans was seeing them pass in the distance in a car.

Almost all cattle on earth encounter humans at a very young age. Many of them are altered in some way by humans. Humans alter cattle visually using brands, ear tags, or ear notching. Humans also alter cattle on a biological level. Many animals receive vaccinations, sometimes antibiotics, in the first year of their lives. We can also change them psychologically. Many dairy cattle learn immediately to have no fear of humans. Beef cattle receive less of this sort of thing. Even so, beef cattle receive some sort of desensitization for humans, horses, dogs, or any combination of the three.

While this bull was a descendant of domestic cattle, its ancestors had lived free from human contact for decades. It would only take one glance from a cattleman to tell this animal from a domesticated one. Working with these animals intrigued me. I am fascinated in the same way that humans are intrigued with uncontacted tribes in the Amazon or with the history of human behavior. I saw for the first time what cattle acted like without humans controlling them. For a young stockman with a desire to better understand livestock, there was no better place to be.

All at once, this animal's life changed drastically. It happened in the blink of an eye; it came together with the scream of an engine. The thundering of hoofbeats and a cloud of dust would make most folks think an army was marching in the distance. It was muscle vs. steel, man vs. beast, brain vs. brawn. The meeting of these two tremendous forces was spectacular. Many believe that in a battle of machine against animal, the machine will always win. The bulls won far more often than one would expect. Many argue that in a battle of the brain against brawn, intelligence will always win. This is even less true than the previous statement.

The cattle living in Kowanyama had been there for generations, evolving through natural selection. They were born and raised on this plot of land and knew where to hide in every patch of trees or field of rocks. The bovines often bested man's technology and intelligence in Cape York. This is one of the wildest places in Australia, where feral bulls are king.

I often get asked why these cattle were free in the first place and why we were catching them. This was not well-explained to me, so I will try to explain it. The land we were catching these bulls on belonged to an Aboriginal community, similar to the land given to the Native Americans in the United States.

The government gave this land to some Aboriginal tribes to live on in repayment for taking the rest of the tribal lands. This is a complicated subject. Most people I spoke with about this were very polarized to one side of the debate or the other. I was hired to work for a contracting crew. To put it simply, the service the company offered was finding and capturing both wild and domestic cattle on large pieces of rural land.

Things were not going well financially for the Aboriginal community. They were not using the land they owned to make enough money. So in an agreement between the bank and the community, the company I worked for was called to clean up the land, giving assistance to corral and capture the wild cattle. Then start fresh, creating a profitable cattle station. The wild adult females were given a brand and released back to the land they came from.

Calves that had grown enough to no longer be drinking milk from their mother were kept in the corral. They sold the calves to feedlots to be grown and eaten for meat when they were of age. The wild bulls were also kept in the corrals. We branded them and then sent the bulls on a road train to the meat works, where they turned into hamburgers. We released purebred Brahman bulls, hoping they would breed the cows on the land. This would increase the quality of the calves produced the following year.

As a company, we also had a crew working on the neighboring stations. These stations were professionally managed, containing only branded purebred white or red Brahman cattle. The three adjoining stations we gathered cattle from totaled somewhere near six million acres. The more common work of gathering, weaning, branding, and vaccinating cattle happened on the two professionally-managed stations.

The wild bull catching mostly happened on the tribal lands. Occasionally, we would find wild bulls on the managed stations as well. I split my time between these two types of mustering. Some men on the crew were better at one rather than the other. The boss distributed the crew where they fit best. They were known as "the mustering crew" and "the bull catching crew." In Australia, to muster is the same as to gather. So, the mustering crew specialized in gathering wild cows and calves using mostly horses, with some help from machines. The bull-catching crew exclusively used machines.

I spent much of my time on the mustering crew on horseback. This crew required horses, helicopters, dirt bikes, quads, and bull catchers. It was a massive undertaking. Each day mustering a new gigantic paddock, each day ending with a new set of cattle in the yards. They kept the boundary fences in exceptionally good condition. This kept cattle from escaping the property and kept the wild, undesirable cattle out. They held the internal fences to a much lower standard. This meant cattle often moved as they pleased, mixing with others without the manager knowing.

We often knew how many cattle were supposed to be in each paddock before starting work that morning. Unfortunately, the numbers they gave us were often very wrong. Some days, we expected 800-900 head of cattle to be in the yards, and we would only get 300-400 head. The following day we would expect 500 head and bring back 2000 head. The last major job was getting the cattle into the "yards" or the working facilities. It was the most dangerous and stressful part of the day.

The process of "yarding up" cattle used every man, machine, and horse available. We had to fit 300-2000 head of cattle into an extensive set of working facilities for sorting and vaccinations. To make this happen, they had to pass through a gate the width of two cars. It was a process of both force and finesse. In old cowboy movies, this is often when you see people trampled to death by cattle in a stampede. This is exactly how it happens modern day. You must apply only enough pressure from the rear and sides to let the cattle know they must go through the gate. Once the first of the herd has started through the gate, the rest of the cattle usually follow.

When we had mobs of cattle up to 2000 head, the process of them filing one or two at a time through a gate could take up to an hour. The trick was applying enough pressure that they kept going into the corrals, not so much that they get scared and run around you or, even worse, over top of you.

I saw more than once how quickly five or ten panicked animals could turn 2000 animals into an avalanche of death. When they ran, it shook the earth. The power of 8000 hooves pounding the ground was frightening. If you fell off your horse or if your horse stumbled and fell, the path of this avalanche would certainly pulverize you. You would resemble a small animal a semi-truck had run over

3
Stampede

On the first day I was working with the mustering crew, I rode a horse named "Gidgy." Gidgy was an old pensioner of a horse that had been there and done that. He was to be my teacher just as much as my mount that day. They warned me that the only thing he did not like was the helicopters, but what horse would? That day, we started in the dark very early. We loaded our horses in the truck and drove them to the paddock we would gather that day.

The sun peeked over the horizon as four or five of us unloaded our horses. We tied water bottles and sandwiches to our saddles, made a simple game plan, and swung up on our horses. As we trotted through the brisk morning air on our fresh mounts, the sun had turned the landscape a lighter blue color. I could hear a distant rumble between the hoof beats. The rumble built into a sound I had only heard in old clips of the Vietnam war. The choppers thundered over top of us, the "THWACK THWACK THWACK" of rotors slicing through the air.

We wore radios strapped to our chests that allowed the men on horses to communicate with the men on machines. Radios also allowed communication to the men in the choppers. The radio crackled to life for the first time early that morning. The voice of a man named Mario echoed through the still trees. He crackled out a typical Australian good morning, followed by the day's game plan laced with the typical Australian cursing.

We had two choppers working with us most days. We gathered paddocks between 30,000 and 50,000 acres in size. Choppers were necessary to spot and move cattle from the air. There simply are not enough hours to cover that much rough land, even using dirt bikes and horses.

We gathered the animals in one group and then walked them in uniform to the corrals. This method is called "coaching." On horseback, we started at the farthest corner of the paddock. We were gathering, and the choppers brought in cows from every direction. In no time, we had 20-30 cows forming a group in a clearing. Once they had come together as a group, we would keep them together using the horses and motorbikes.

Cattle are much easier to handle when they have come into a group. Even a group of three is enough to make the cattle calmer. They move as one. We would then drive our mob of cattle using the horses. We drove the cattle toward the shortest and easiest route to the corrals we aimed to put them in that night. Through-out the day, the choppers and motorbikes continued to bring cattle out of the brush. The machines added cattle into the mob we had formed. As the day went on, the mob got larger and larger.

Around midday, we stopped and took a break for lunch at a water hole. The cattle would have a drink from the water hole. These water holes almost always contained crocodiles. The horses sucked down a belly full of water as well. We ate our sandwiches from our saddlebags. At least once every week or two, the cook forced us to make pickled tongue sandwiches, so we wasted no meat.

Cattle, horses, and people took the free hour to relax in the shade and escape the deadly Australian sun. It was not uncommon for the cattle drinking at mid-day to get snapped up by a crocodile and never be seen again. We would hear a second or two of splashing, and the animal was gone below the water's surface. This kept all humans and animals on edge around the water.

After the lunch break, we started the cattle up again into a lumbering mob of tired beasts. First, they gave us the cry of a teenager waking up for school. Then, their eyes asked, "Just five more minutes, please?" Both stopping the cattle and getting them going was a long and tedious process, a process we tried to avoid but once a day for lunch.

We had reached a point where we were on a large hard-packed country road, a big one that road trains frequently sped down. Road trains are huge semis pulling three or four trailers. From here back to the corrals should be smooth sailing, they told me. A sense of relief came over both the people and the animals. I rubbed the inside of my knees raw from my first day in the saddle, a type of saddle that I had never ridden before.

By this point, some of my coworkers had gathered the courage to ride over to me and bull shit a bit. They asked where I was from and what it was like. One young ringer rode over to me, maybe 18 years old, and started asking me about the chewing tobacco I kept in my lip, a bad habit rarely seen outside of the United States. He stopped mid-sentence. Everyone felt it, but nobody knew what it was. The cattle had shifted; they were uneasy. The Aboriginal people are firm believers that spirits still inhabit these wild parts of Australia. They would later say the spirits caused the chaos about to unfold.

In the blink of an eye, the cattle at the front of the massive mob of cattle had stopped, whirled around 180 degrees, and started pushing back into the herd. This slight change by a few animals set off a chain reaction. The entire mob had stopped and spun around in a hurry. I was in the back riding drag, where the new guys ride. Before I knew it, I was the only thing standing between a couple thousand cattle and open country.

It was my first day. I committed a classic cowboy mistake by trying to hold my ground too long. I flailed my arms and yelled at the top of my lungs to deter the cattle from trying to run past me. My effort was useless. By this time, the helicopters hovering above had realized what was happening and swooped down incredibly low to the ground to assist.

Since I was now in the lead of this runaway train of death, the choppers dropped in directly above me. The chopper swirled dust into my nose and eyes as I sat in the rotor wash. As promised, my horse Gidgy let me know how much he disliked our situation.

My horse began bucking with the chopper just above my head and wild cattle stampeding around us. As he leaped high into the air, I made my peace. I was going to die. I leaned forward and hugged his neck to stay on his back, trying to stay as low as possible while the chopper was directly above us. I had visions of him throwing me straight up into the blades, then chopping me into tiny pieces and being left there for the birds to clean up.

Finally, we were out from underneath the chopper but still trapped in the middle of the cattle. Only then did it occur to me that everyone else had gotten out of the way. Now they were screaming at me to get the hell out of there before the herd killed me. We ran with the cattle at full gallop, trying to fit through every small opening that presented itself, moving us slightly closer to the edge of the herd and safety. We rode for what seemed like an eternity. Finally, my horse and I emerged from the chaos panting, sucking in dust, covered in sweat, and thankful to be alive.

Eventually, we got the stampede slowed and then stopped. The boss told us he thought we had only lost a couple of hundred head in the scary ordeal. We got the cattle started in the right direction again and kept them moving all the way to the corral. We stuck the cattle in the corral as the sun sank low in the sky. I stepped off my horse finally after 11 hours in the saddle. My legs shook with exhaustion.

The newly-formed wounds from the rubbing of the saddle stung, and my eyes burned from the dust that filled them. I had become nearly unrecognizable with the thick layer of dust coating every inch of me. I had never been so excited to climb into my cold, smelly bedroll. 4:30 am would come early the next morning, and we would do it all again.

4
Green Ants

In the months I lived in Australia, I slept inside a building for only four or five nights. I spent months at a time living in a "swag." At first, I slept on the ground, an obvious rookie mistake that the seasoned veterans laughed at. I did not have the money or even the ability to buy a cot to keep me up off the ground. After seeing far too many snakes in my first week in Australia, I decided I must get up off the ground, an attempt to keep one of Australia's many deadly snakes from crawling in bed with me.

My Mom had given me a gift before I had left: a hammock. I concluded Australians must have never heard of them. Nobody used a hammock; they are much easier to use than any cot. The first night I tied my hammock up between two trees. I hung it before dark, on the only trees in sight that were the correct distance apart to hang a hammock. Everyone stared at me. To my knowledge, I was the first American to be in this part of the world doing this. They often stared at me. As I climbed into my hammock, I felt the first sense of relief I had felt since landing in Australia. I was finally relaxed in a safe bed.

Not long after I had dozed off to sleep, I woke to a sharp pain in my neck. Initially, it almost seemed like a nightmare. Then a sharp pain in my stomach woke me fully. Immediately, I had the sensation of things crawling all over me. It was pitch black other than the stars. I did not know what was going on. Then another sharp stinging pain ripped up my leg. I realized I was covered in some sort of bug that was trying to kill me.

I flopped out of my hammock onto the ground, still in my swag, with a thud and a quiet yelp. I was being bitten all over my body. It was excruciating. This was much more than a bee sting. I frantically flailed out of my swag, rolling around the flat, trying to get my clothes off.

My first thought was to run to the pond we were camped next to. I had seen people in the movies get away from attacking insects by diving under the water. The problem was that I was in far north Queensland, and every water hole had a few crocs. To go splashing into the water after dark was suicide. I stripped off most of my clothes and rolled around in a panic, trying to brush off every inch of my body. When even that failed, I did the only thing I could think to do. I just ran. I bolted across the open land until finally, I hit something on the ground. I tripped and rolled across the sand.

I looked back to see a light come on. I tripped over one of the Aboriginal men working with us. His name was Frost. Frost had thrown his swag on the ground. He looked up at me with a groan and said, "The fuck mate?" and just then, I got another bite which reminded me why I was moving. I yelled, and he shined his light on where the bite had happened. I was being attacked by green ants, something I had never seen. Frost scolded me to stand up straight as he brushed the remaining ants off me.

Frost and I then walked back over to my hammock and the swag on the ground next to it. Frost shined his light on it to reveal that the hammock appeared to be moving with how many ants were swarming it. The swag had much fewer ants on it. We shook it out, and Frost informed me of how stupid I was. He told me to sleep on the ground near him.

The next morning when I woke up at daylight, I walked over to my hammock. I collected the things that I had scattered in my late-night panic. There was not an ant in sight. Had I not tripped over Frost in the middle of the night, I might not have believed it had happened. Since I saw no ants in the daylight, I wondered if it was all a nightmare, possibly the spirits that Aboriginal people spoke of so often.

That day, they informed me that green ants are one of the most feared animals in the area. Green ants have the most painful insect bite I or anyone in the crew had ever felt. I learned they make nests in the trees made from leaves. The ants glue them together to form something that looks like paper mache. I had a few other run-ins with them in Australia when catching bulls. We occasionally hit trees while driving at high speed on a motorbike or in a buggy. Some of these trees had nests in them, and I would knock the nest down on myself.

It was one of the many lessons I would learn the hard way. One of the most feared animals on Cape York: ants. The next night, Archie had woken to the sound of his dog in a fight. When he shined his light, his dog was whipping a very poisonous snake around. He had killed it. That snake was awfully close to where I had just run barefoot in a panic, trying to escape the green ants.

Having figured out that, I could not lay on the ground safely. I knew I could not hang from the trees safely to sleep. So I began looking for materials to build myself a bed frame to stay up off the ground. A few days later, I pulled over while passing an abandoned building. I found an old rusty bed frame and brought it back with me.

I then found a few empty fuel drums, laid them on their side, and set the bed frame on top. I had effectively made myself a bush bed. It almost made my coworkers proud to see I had shown some true Aussie innovation and made myself a bed. It was ugly, but it did what I needed it to, and finally, I could sleep.

About a week later, I woke to the sound of large raindrops hitting the top of my swag. I was so tired I just fell back to sleep. When I woke up in the morning, I realized what I had heard in the night was not raindrops. It was something much worse. I had placed my homemade bed under a tree for protection. That night, hundreds of giant fruit bats landed in that tree and began shitting all over me. Everyone laughed at my expense that morning, but I could not care less. At least I had slept.

5
We're in the Spirit World Now

The crew I worked with in Australia is one that I will never forget. The group of people would shock most anyone at first. It was a crew made up of folks that society forgot, or maybe they told society to leave them alone. When you hear the statement "I was born in the wrong generation," these are the people that truly acted upon that statement. Not one of them would be considered "normal" if they went into one of the big cities in Australia.

I worked with drug addicts, alcoholics, schizophrenics, renegades, outlaws, rogues, legends, rodeo champions, religious folks, atheists, men, women, old, young, wild ones, quiet ones, black fellas, and white fellas. Regardless of their backgrounds, they were some of the roughest, toughest people I have ever met. People who did things the hard way. The old fashioned way, the way they believed to be correct. That was all that mattered.

Being around people like this was a major paradigm shift for me. It was like I had found my little tribe of people proud to be outcasts. We were the lost boys. We were wild. We were strong. We were free. It was liberating. This realization was life-changing for me. I realized that I felt most comfortable with those who lived on the lunatic fringe of society. I was truly at home where the wild things are.

The people I worked with often referred to themselves as "ringers from the Top End." A ringer is an Australian cowboy or stockman. They would meet you with fierce opposition if you called them anything but that. I was told that the term came from the fact that to stop wild stampeding cattle, you must ring them up. Ringers must turn them and make them run in a circle or ring until they have tired enough to stop running.

It was also a huge point of pride to tell people we worked up on the "Top End." The Top End refers to the north end of the province of Queensland. It is also known as Cape York, one of the last areas of Australia settled by colonizers. The Top End was still full of native Aboriginal people with little infrastructure in place. We told people we were Ringers from the Top End, and instantly we were admired, seen as the real deal. I saw places in Australia that most Australians never see.

Possibly one of the most unique aspects of my time in Australia was my chance to live and work with Aboriginal people, fascinating people who have fought for generations to maintain their freedom. They are the last people representing what Australia used to be. Now, they are forced to live off government programs and are pushed inland and away from major civilizations. They fiercely protected family, homelands, tradition, ancestors, and spirits. It was shocking to learn that the community I lived in had almost lost its native language. Only the oldest tribe members could remember their native tongue.

Every Aboriginal person I met was dreadfully afraid of upsetting spirits. They spoke of their dead ancestors as if they were close friends. They took their hats off and quickly conversed with every grave they walked past. As a sign of respect, they always stepped down from their horses. Aboriginal crew members often would tell us of their dreams the night before, dreams of their ancestor paying them a visit and telling them of things that would happen soon.

It was spine-tingling when some of these dreams came to fruition. The spirit world was not something to be played with on Cape York. I heard stories of entire crews quitting and leaving in the night. Some stories told of the Aboriginal crew members having the same terrible dream, and they all left together. I also heard stories of some of the white crew members trying to play pranks on Aboriginal people in the area. Pranks they did not find funny, causing them to leave, fearing for their lives.

One of the most commonly talked about spirits is Feather Foot. I was told by my friends that he is an Aboriginal Grim Reaper. He comes in the night to collect your soul. He would leave a warning if you were on a path to meet your death before your time. Feather Foot would do this by leaving a feather on your pillow or sometimes near your bed. If you awoke to this, you should immediately leave the area because death is near.

The men I lived with loved to play tricks on anyone they could, being the rowdy, rough group of men they were. Some pranks were distasteful and even quite dangerous. However, it made it even funnier whenever the person made it out unharmed. One of those distasteful pranks I heard of someone playing was aimed at the Aboriginal men on the crew. One man slipped a feather onto a man's pillow in the night.

When the man awoke, he was gone in a flash, all the other Aboriginal men on the crew following behind him, even though not all the men had not awoken to feathers on their pillows. It was enough of a bad omen to cause them to all quit their jobs on the spot and spend a few days walking back into town. The boss was not pleased when he learned about the prank that caused most of his crew to depart, leaving them shorthanded until they could find additional help.

In one of our camps, we set up our portable kitchen and dining area under a large tree to protect us from the sun. Under this tree was an old house that had fallen, and next to it stood an exceptionally large termite mound. Termite mounds are rock hard. The structures stay standing long after the termites abandon them. The mounds are scattered all over the flats of Cape York. This mound was special.

One elder came to visit us one day, claiming a baby was inside that mound next to where we ate and cooked. The elder said that if we altered the mound, we would all be cursed. While this seemed absurd to most of the crew members. The crew quietly respected the elders' claims. One time, a white crew member did nothing more than touch the mound with his hand. The touch nearly caused all our Aboriginal crew members to leave for fear of being cursed. It took some smooth talking by our boss to convince them they were safe to stay with us and continue helping us work.

7
Bush Tucker

Living somewhere this remote meant that trips to the supermarket rarely ever happened. We had no clean water, the meat we ate we killed, and all the bread and baked things were handmade. Occasionally we would get a delivery of eggs and fresh fruits and veggies. Still, mostly we ate what we could produce ourselves. The bread had a hard crust that protected and preserved it from the outside world. It was a classic bushman bread. The Australians called it "damper."

We could carry damper in our saddlebags, which would keep for days. When it came time to eat it, we had to have a sharp enough knife to cut through the tough outer shell. They referred to all the food we ate as "Bush tucker." Bush tucker is food that can be prepared and carried around in the bush without refrigeration.

We would kill a fat beef cow every two or three weeks. We would kill the fattest one we could find. This would ensure two things; fat cattle taste better and are usually healthier. The other reason we killed fat cows is that they were not raising a calf. Good mother cows often give their calves most of the nutrients they consume. They do this through their milk. The best mothers to the untrained eye are often the worst looking. Fat cows rarely raise a calf, and if they are, they probably are hardly giving it any milk.

After a precisely placed shot with a rifle, one shot, one kill, we cut up the animal. This is a process that dates back ages, one of our most primordial processes of harvesting meat. It was hard work, but it was something that everyone enjoyed. There was excitement about fresh food. Something is rewarding about the process of bringing raw food full circle to the table. We often only had a few sharp knives, a rifle, an ax, and maybe a saw if we were lucky. There are no food safety protocols here. It is men cutting, pulling, and sawing anyway they can get the job done.

In most places, they would do this in an exceptionally clean area indoors. Not with our crew. We did it right where the animal fell, out in the bush. It is also common modern-day practice to hang the animal up by its back feet. This helps with the draining of blood and the removal of internals. It keeps the animal off the ground once you have started cutting so the meat cannot be contaminated. We did the entire process while the animal was lying on its side in the dust, covered in flies. Many of my best friends are in the meat industry. I took a few courses at University discussing the science and safety of breaking down meat. I knew immediately that this would never fly anywhere but in the bush.

I was told to gather an armload of tea tree branches immediately after the animal died. We used them for two different purposes. The first was to use them as a somewhat clean place to set meat. Once the men started with their knives, they just started chucking pieces of meat in the back of our ute. That area was full of oil, gasoline, mud, dust, and many other things we did not want to touch our food.

We laid down a thick layer of tea tree branches so the meat would rest on top of the freshly broken branches. The second use was to clean the dead animal's hide. With a tree branch, a couple of men would start brushing the hide and hitting it like a dirty old rug. This was to remove all flies and as much dust as possible from the animal before cutting open the hide.

Once the animal was opened down the belly, they pulled the skin back like an orange peel, exposing the muscles we needed to carefully remove. They removed the large muscle groups one at a time. Then the crew picked out the smaller pieces they considered delicacies. They always took the tongue out. A saw or ax took the ribs out. Then everything that could be taken from that side was removed. We folded the skin back over the animal and flipped it to the side that had the remaining meat. They did the same thing on this side, and that is when things got different.

All the guts of the animal remained inside still; this was everyone's favorite part of the animal, it seemed. We then removed the liver, kidneys, lymph glands, heart, and small intestine. We would dine on this part of the animal the very night we killed it. These parts did not keep for awfully long, and there was something very traditional about doing this. When we left the animal, we left only pieces of the head, backbone, and back leg bones, all cleaned free of meat. We wasted nothing.

When we returned to camp with our cuts of meat, we hung the meat in one of our metal buildings. We did this to keep as much dust as possible off the meat. To hang the meat, we used wire to make hooks. This wire was the same wire we used to fix fences with, or tie together broken pieces in the corrals and even repair broken car parts. The meat would stay on these hooks outside for a few days. This would allow the meat to cool down and cure a little.

The meat could only cool at night because, during the day, the meat hung in a metal building. The sides of that metal building were often so hot from the sun that you could not touch them. So, after a few days, we transferred the meat into a massive cold room our boss had fabricated in the front of a semi-truck trailer. It had a diesel generator on the front that kept the room cold.

The night of a kill was a feast. We ate massive rib bones from the cow. The heart was cut up into steaks and cooked over our fire. We dropped the liver and kidneys into hot oil and cooked them up one chunk at a time. One of my favorite foods, along with everyone else, was the curly gut. The curly gut is the small intestine. We would pull a meter or two of the intestine out and then clean it, rinsing everything from inside the tube-like structure out. What it contained was just a few hours from being cow poop had the cow not died. The intestine was then cut up into bite-size pieces. We cooked it in oil, then we sprinkled some salt on it, which tasted fantastic.

Once, when we had a less experienced cook, we all bit into our fresh pieces of the curly gut we had just harvested to find blended spinach-like filling. The cook had forgotten the part where you wash the intestine before cooking. It was like we had cow-shit-stuffed calamari, if you can imagine that, without gagging.

Growing up in the mountains in Colorado and Wyoming, where water is plentiful, I never thought about its scarcity much until Australia. We needed to carry enough water with us or make camps close enough to water that we could collect it. We did not use filters, chemicals, or boiling to clean our water. Instead, we drank it straight from where we found it.

This sometimes meant we drank from old, abandoned wells. Other times we drank from manmade reservoirs, the same ones the cattle drank from. Because the water was so silty and full of sand, we often had to pump the water into one tank, let the sand settle to the bottom, and then pump the water to another tank to consume.

We had a water tank big enough for two or three men to bathe in at one camp. This was our drinking and washing water. We often added cordial to our water when we had the option. Cordial is a very concentrated sugary liquid with a lot of flavor. When cordial is added to the water, it covers up the taste of dirt. One day, after running out of cordial, the boss said something was wrong with the water.

Everyone assured him he had just liked having a sugary, fruity flavor added to his water. He refused to believe that. He walked to the drinking water tank, unscrewed the cap, and looked down inside. The boss was right. There was something very wrong. We had all been drinking it for days or weeks and masking the flavor. On top of the water floated two dead cane toads, halfway deteriorated by now, rotting away in the water we all slurped down unknowingly.

Water was incredibly important out in the bush when we were bull catching and mustering. I seemed to run out daily because I did not own a big enough water bottle. The bottle needed to be small enough that I could carry it on my horse. A man could easily drink a gallon of water daily, if not more, out there in the blistering sun.

I carried only a one-liter bottle with me. My coworkers repeatedly scolded me for trying to bum water off them. It was not until then that I understood what thirst was. I felt I could fall off my horse at any minute from heatstroke, but I never did. When we came close to a reservoir, we all filled our bottles with opaque, silt-filled water. We drank it like we may never drink again.

Getting a drink of water in this part of Australia is harder than you would think. The bigger the body of water, the bigger the crocs that lived there, so nobody ever ventured up to the water's edge. We stayed away because before you knew what was going on, a six-meter saltwater croc could drag you to your watery grave. We just tied a string to fill water bottles, threw them out into the deeper water, and then pulled them back in.

The horses needed water often, like the cattle. So, when we stopped to get them water, we simply got off our horses and let him walk to the water and figure it out. The horses were so thirsty they would not go anywhere but straight to the water. We did not sit on our horse when he drank because if a croc grabbed the horse, you were going in together.

8
Bull Catching Machines

The act of physically capturing a bull is a skill many will never understand. We couldn't act simply out of fear, anger, or enjoyment. The slightest miscalculation could mean severe injury. Being trampled or gored in a place as remote as Cape York could mean death.

There were many ways to capture the bulls, some more dangerous than others. As with all adrenaline junkies, the most dangerous method was the most fun and preferred. There is something blissfully frightening about a person who can simply look death in the eye and smile. This occupation was just that, playing with ticking time bombs. After defusing one and having a laugh with your mates, you'd get a sip of water and go as quickly as possible to find another.

The old-fashioned way, you might call it the original way, was to pull the bull down by his tail. This did not mean springing out of the bush, surprising the bull, and just grabbing his tail.

That sort of thing is not possible with wild bulls. It all started with horses. Horses can run much farther and faster than cattle. A man on horseback would pick out a bull, and the chase was on. He would chase the bull at top speed on his horse, crashing through brush, termite mounds, washouts, and whatever else may come across his path.

When the bull became tired, his back end would be the first thing to weaken. While riding through hell or high water, the ringer had to be very observant of the bull and ready to pounce at any second. When the bull had slowed to a trot, the ringer would watch the rhythm of his back legs. Occasionally the bull would stumble. Sometimes he would hop on one leg, skipping a beat, trying to give it a rest. The most obvious sign he was tired was that he stopped. When the bull slowed enough to catch, the ringer needed to quickly dismount his horse and get ready to fight for his life.

Once the bull began to skip or stumble, he would be slow enough that a light-footed human could run up behind him and get his tail. If the bull turned around ready to fight, the ringer had to be ready to dodge his initial charge and grab his tail, hoping to avoid his horns. Regardless of how the ringer got their hands on the bull's tail, the ringer had to immediately wrap the brush part at the end of the bull's tail around their hand for the strongest possible grip. Now the fight was on. The ringer would grab the tail and hold on for dear life.

Using their body weight, the ringer would pull in a direction with all their might, watching closely to see when the bull might cross his legs or stumble, which might let them pull the bull over. It is a matter of life or death that the ringer does not lose their grip on the bull's tail. By maintaining this, the bull physically cannot reach the person with his horns. The bull will take the ringer for one hell of a ride, being dragged around the bush through sharp barbed plants and uneven hard ground. If they hold the tail, at least they have taken his horns and ability to run them over out of the equation.

It is a monumental task to pull down a bull. It takes all the strength and endurance a man can muster. Sometimes the bull has more endurance than the ringer, and this is where things get even more dangerous. At this point, ringers lack the speed or ability to get away, so letting go is not an option. In the worst-case scenario, the ringer must sacrifice his body to get the bull to tumble. This was the riskiest method, and I only saw it work once or twice.

Swinging on the bull's tail like Tarzan on a vine, a ringer swings out directly beside the bull and swings into the bull's legs. If the ringer holds his legs wide open in a V shape, he has a chance at catching the bull's back legs. He can then hook his legs together and squeeze for all he is worth, hoping that holding his back legs together will cause the bull to fall. If it does, he has to clear out of the way quickly as the bull hits the ground.

Whatever method we used to get the bull down, once he was on his side, we had to work with haste to keep him there. The bull is seven times bigger than the person catching him, so jumping on top of the bull will not hold it down. The only way to keep him down is to keep his feet off the ground, all four of them. This is done by pulling as hard as possible on one of the back legs to get all his legs pointing up in the air, like a turtle upside down. Then, we would take off our first belt and tie up the back two feet in that position. This was difficult to do alone, so having another person there could make things much safer.

Next, we worked on the front end of the bull to get his front feet tied with a second leather belt. You have to be even more careful working on the front end because the bull is incredibly hard to control from the front. Bulls are deadly accurate with the tips of their sharp horns. We would cover the bull's eyes, applying light pressure. Then he could neither see to hit us nor raise his head to get up. Once we had him where we wanted him we would quickly tie the front feet with the second belt.

Once the bull was tied up, we had a few options. The first and quickest was the bull truck, but that only worked if there happened to be one nearby. The bull truck could pull up, load the bull, and then move on to pick up other bulls the crew had caught. With the bull lying on his side, we would hook a winch to his horns and slowly pull him up into the truck. A bull's horns are so solid that they served as a good anchor and wouldn't cause him any harm.

If there was no bull truck nearby, we would need to get him down on the ground near something we could tie him to. Usually, we would try to trip the bull down near a solid tree. If we could get him close enough to a tree, we could take a thick cotton rope and tie one end to the bull's horns and the other end to the tree. The bull could stay tied to this tree for a few hours before the bull truck would come get him. There would be enough food around the tree to keep him happy.

If there was no bull truck or tree nearby, we could use a bull catcher. Once the bull was down, we could tie a rope to his horns and tie the other end to the armor plating of the bull catcher, which was usually some old modified Toyota. Then the bull catcher could drive slowly, leading him to the nearest tree, and we would tie him to that until a bull truck could come get him.

When running around in the bush chasing bulls, we always carried our two leather belts for strapping the bull's legs. We hooked the belts loosely around our chests for easy access. We also always carried a radio in a holster on our chest. This was our lifeline and only chance for help if we got in a bind. Last, we carried a coiled-up piece of rope, like the rope American cowboys use in rodeos but much softer. That rope could be an incredibly handy tool. The Australians used the rope differently than the American cowboy or the gauchos of Argentina.

They did not tie the rope to anything before throwing it and catching the bull. When they threw the loop around the bull's horns, they needed to immediately run to the nearest tree capable of holding the bull. They had to wrap the rope around the tree a few times as quickly as possible, using the friction of the rope on the tree to their advantage, trying to hold the bull. The bull would fight and pull. The second he gave any slack, they would pull up the slack until he was close enough to the tree they could tie him off and walk away.

There were other options for catching the bulls that were slightly less hazardous to human life. These were mostly motorized options, and are probably the most common methods used now. We often used motorbikes instead of horses. We could chase bulls the same way you would on horses, but instead of letting our horse go, we would drop the bike on its side. We then would hit the ground at a full run to catch the bull's tail. We also sometimes caught the bulls without getting off the motorbike, but it was difficult.

Riding up on the right-hand side of the bull, we would let go of the handlebar with our left hand, grab the tail, and wrap it around our hand. Then, with one fluid movement, we would twist the throttle and turn slightly to the right. Turning to the right gave us the best chance of avoiding the bull's horns slashing our leg. We then pulled the bull's back end slightly sideways, causing his back feet to cross, and then stumble. Usually, we had someone riding directly behind the bull on another motorbike. They were ready to jump on the bull and secure him once we had tripped him. Then, once on the ground, we tied him up.

The quad bikes or four-wheelers used in bull catching have the same sort of plating that the bull buggies do. However, they are lighter and faster, so they can catch bulls easier in most situations. With the quads, it is usually less of a game of precision and more of a game of power. Using the massive front and side bumpers, we would drive up next to the bulls, and much like a police car, we pulled the PIT maneuver, pushing on the back end as the bull ran until we spun him out, causing him to trip and fall.

The nice thing about the quad was that we didn't need to jump off and wrestle with the bull to keep his legs off the ground. Instead, we could simply park the bumper against his belly. It had the same effect, holding him up, and then we could get off and strap the bull all by ourselves, not needing someone to hold him down. We used the bull catchers in the same way. They were just much slower and harder to maneuver. Before quads became so fast and powerful, bull catchers were the ideal catching machine. Now they have almost become outdated in this sense.

Have no fear. The bull catchers have not lost their place in the bush just yet. The last method of catching bulls requires the burly old bull catcher fitted with newer technology. This is the one method I did not get to experience while in Australia, but it has become quite popular recently. It is a method used in the old John Wayne movie *Hatari*.

They have fabricated an arm that sits on the front driver's side corner of the bull catcher. They can hydraulically operate the arm so with the push of a button, the arm drops around the animal's neck, keeping him from escaping. With the arm, they drive up alongside the animal. Often at a full sprint, they can position the bull catcher precisely so that the arm can close and clamp around the bull's neck. Then the animal can not pull its head out. The bull can be walked slowly to a tree, tied up with the soft rope, and they can drive off to catch another.

9
Cow Dogs

One of my favorite things to observe while traveling is how people use animals to aid them in their daily lives. Across the board, two animals accompany meat-producing livestock across the world: horses and dogs. The horses are all used for the same thing, riding. While the riding style and equipment are vastly different, their general purpose is the same: transport humans faster and farther than they could move on foot.

Dogs are a different story. In Australia, the way people use dogs differs from everywhere else I have traveled. The Australians had two main breeds where I was working. They used kelpies to move stock, but they also used hanging dogs.

A hanging dog is not trained to push the livestock like you see border collies and kelpies do. The purpose of the hanging dog is to work from the front of the cattle, slowing or completely stopping the bovines. Catching the bulls the way we did, it was always easier to have a partner. The partner would get chased by the bull as you pulled its tail, making the job slightly easier.

When ringers had a hanging dog, they did not need a partner. Cattle instinctually will go after dogs before people most of the time. So, if they released their hanging dog to distract the bull, they would have a much better chance of grabbing the bull's tail without being noticed.

Pit bulls are illegal in the part of Australia where I was living. We often got stopped by the police when they saw our dogs. The police asked what breed they were. The answer was always simply, "I don't know. He's a mutt," and the police could not prove otherwise. Most hanging dogs were of pit bull blood or another breed called a Stafford terrier.

The terriers were like miniature pit bulls and tough as nails. They instinctively went to the head rather than the tail of the cattle. They would get out of the way when the owner told them to. It was incredible how well they listened. The dog would simply disappear when the owner yelled his command in the heat of battle between bull, man, and dog.

The dogs were usually dual-purpose; they were also used for hunting pigs. I was told that pigs were free game in Australia: shoot as many as possible because they are invasive and cause massive crop and habitat damage. There were also big bragging rights for the person who killed the biggest pig. So we often would go after work with the dogs and catch pigs. The dogs would be "plated up," meaning we would put vests on them that protected their necks and stomachs. The protection was crucial because the pigs had large tusks that could easily kill a dog if unprotected.

The kelpie dogs worked fantastically; they were very skinny, like a dingo, perfect in the Australian dust and heat. They rarely used kelpies while mustering, mostly because it would require them to cover huge amounts of land. The distance would likely kill them. We used them around the station. One of the dog's biggest jobs was to "educate" the young cattle.

The young cattle became quite wild once removed from their mothers. Trying to drive them in a group was like herding cats. They ran in every direction once you opened the gate. The method they came up with to break them of this habit was called "educating." It was a two-step process that often took an entire day, or sometimes more.

The first step was to let the dogs teach them to stay in a herd. The kelpie dogs would go after any of the weaners that stuck their heads out of the mob. If something tried to run away from the mob, the dogs would team up. If the cattle stayed in a tight group, the dogs simply ran in slow circles around them, keeping them that way. Often, one brave weaner would think it could escape by breaking free of others and running away. The cattle had been indirectly naturally-selected for this trait. Any cattle that ran in the opposite direction as the herd would often not be caught for years, maybe never caught at all. When they did this in the heavy wooden corrals at the station, they could not escape. So, the dogs educated the young cattle, who learned they were safer when they returned to the group. This process would sometimes go on for hours until the stockmen felt the cattle were ready for step two.

Step two involved letting the cattle into a small paddock with sturdy fences. Using horses and dirt bikes, we would walk the cattle around for a few hours, teaching them to stay together. It taught them to walk instead of running. This was often a wild practice. It was horses, dirt bikes, and dogs flying around at full speed for a few hours until the cattle had calmed and learned to walk together.

We were never taking the cattle anywhere specific, simply walking them so they'd learn to walk. The results were really impressive after a day or two of working like this. Only a couple of days before, these wild young cattle had tried to jump out of the corrals. They had tried to run over people and dogs and were quite spooky. However, they quickly turned into respectful, calm cattle that were enjoyable to work with. On stations that ran tens of thousands of head, it was nice to see them take the time to work with their cattle and train them. While it was not much time, it was more than even small ranchers do in most other parts of the world.

The last common kind of dog was almost a cross between the other two types of dog. They were often referred to as stump tails. Many people asked me upon my return if people used a lot of blue or red heelers in Australia, and I never saw one. Stump tail dogs were the closest I saw to a heeler. They had the thick head and body of the pit bull, but they were thinner and could cover more distance. Their faces had very thick, coarse hair that felt wiry. They could work on both ends. They could both go to the nose like the pit bulls and push from the back like the kelpie.

10
The Burnt Land

I worked with a man named Brendan, who was raised on the Cape. I heard some people echo that if he kept up what he was doing, he would die a bull-catching legend. His father was somewhat of a legend. He had spent decades on the Cape, working his way up to a station manager position. Brendan was always chasing his Dad's record and doing it himself. He lived in a trailer he pulled behind his ute. The trailer had housing and a place to haul horses or a 4-wheeler.

Brendan also owned his own Mad Max style bull catcher he used almost daily. He would disappear for a whole day on his own and come back and say, "I caught ten bulls today. I pinned their location on the GPS." That's it. Ten bulls would have been a tremendous day for any other man at camp, probably a record for one day, but Brendan would do it often.

He claimed his record for bulls caught was somewhere in the high forties, maybe low fifties. His Dad, decades before, had caught over sixty bulls alone in one day. That was the number Brendan was trying to beat. I saw neither feat take place. But, after seeing how easily he would catch ten or fifteen in a day, I could imagine it was possible if the conditions were perfect.

We often went out bull catching as a group. Sometimes I would get to ride a Honda 230, or sometimes I rode in the bull catcher as the "strapper." When the bull catcher driver would get a bull knocked down, the strapper went to work. It was the strapper's job to jump out as quick as possible and strap the feet together. It was hard and terrifying work.

One day, we assembled the entire crew and went to catch bulls, at least ten men mounted on machines on the prowl. The day started well. I watched one man tip a bull by grabbing its tail while riding the motorbike at a dead run. Then, another man drove his dirt bike into a termite mound buried in the grass. The collision flipped him over the front of the bike and broke his shoulder. We got him sorted out.

Another man was chasing a bull later that day when he realized at the last second that there was a barbed wire fence in front of him. It was in poor shape, so it was hard to see. He hit it with tremendous force. The barbs ripped through his clothing and skin as he slid to a stop. It was as if someone had tried to cut him with a chainsaw. He rolled around in agony, but he gritted it out, got back on his bike, kick-started it, and rode the rest of the day.

We all gathered back up to eat "smoko," Australia's favorite meal. It's like teatime or an early lunch. We made a small fire and ate our pickled cow tongue sandwiches. Nobody really enjoyed the pickled cow tongue. Everyone pulled out their "billys." A billy is a small pot capable of holding a quart of water. They set them next to the fire and boiled water to make tea or coffee. We made a loose plan to all go out to burned land. Aboriginal people often lit massive grass fires that burned for weeks to rejuvenate the land. When they burned the tall grass, it removed all the hiding places for cattle.

The burned land was covered in ash. As you rode across it, it made enormous clouds of black dust. The dust would turn every bit of our exposed skin black. From a distance, it was hard to tell who was Aboriginal and who was not. We converged on a big opening as a group and found a large herd of bulls. We all rocketed out of the bush simultaneously, surprising them.

Bulls and bikes flew through the thick dust in every direction. I was helping push one bull toward a man on a four-wheeler so he could catch him. Through the dust, we lost him, so I turned to help another guy nearby. Earlier in the day, I crashed my bike and broke the shifter. It still worked but was hard to use.

I turned to my right, and Brendan ran a bull straight at me. A man was on foot, getting ready to trip the bull. I realized I needed to move immediately, but in my panic, I could not work the broken shifter. The bull plowed directly into me. I jumped out of the way at the last second, but he hammered my bike, tripped, and fell on top of it.

We got all the bulls in that clearing captured and moved to a second clearing where we found another herd of bulls. The large logs on the ground still smoked from their internal smoldering. Ash covered every man, and the smell of grassfire was still strong. The men nearly blended into the earth. Then, after drinking water, washing the ash from our eyes, and catching our breath, we surprised the next herd of bulls.

Again, a whirlwind of motorbikes, men, and bulls created a massive dust cloud. We had three bulls lined out simultaneously, three four-wheelers pushing the bulls, and three motorbikes following. This meant one bull was mine to trip all alone. I tried and tried to get into position to grab the bull.

The combination of my fear and lack of experience stole my chance at glory. I could not get the tail in time. Another man had to jump in and get him for me. I felt incredibly low. Rookies get plenty of chances to chase bulls and allow the experienced guys to trip them down. Rarely do rookies get served a bull on a platter like they gave me that day, and I missed my chance.

After a long day, we had caught nearly thirty bulls and tied them all to trees. We would return early the next morning, before the sun was even up, to load them on the truck. The sun was low in the sky, turning brilliant colors. It was time to race home. Most of our machines had no lights. I rolled a cigarette, still feeling bad for myself after missing that bull. I turned and headed home, riding side by side with a seasoned ringer from our crew named Spook.

Spook and I were leading the pack back home, like a band of heathens aboard our machines, black as the night. The sky had turned a beautiful orange and purple when a bull ran right in front of Spook and me. He had massive horns. We slowed down. Spook smiled a nasty black ash-filled smile and took off after the bull. I could not let him go alone. I twisted the throttle and raced after them.

Spook got close, crashed his still-running bike to the ground, and grabbed the bull's tail. He was in a bad spot. There were brush piles and barbed wire fence awfully close. Both things could hurt him if he did not trip the bull quickly. I dropped my battered old bike to the ground and ran right at the bull's head, drawing his attention. The best chance to trip the bull down was when I let the bull chase me. I began to tire, and the bull nearly caught me.

The bull swung his long, sharp horns wildly. I avoided them only by throwing my hands up and sucking in my belly. If he had caught me like this, I surely would have been holding my intestines in my hands a few seconds later. He came so close with this swipe he hooked my shirt, tearing it open and putting a big hole in it.

Spook tripped the bull, we strapped him quickly, and the other guys arrived to help us get him to a tree and tie him up. I wondered at that moment why the hell I was even here. Adrenaline pulsed through my veins as I remembered how I nearly watched myself get gutted like a fish. I was shaking uncontrollably. Spook looked up at me and said, "Texas, that was fucking awesome mate! I wouldn't have got him if you didn't help me," He reached his hand up to give me a high five.

As our ashy, black, and callused-covered hands smacked together in the low light, I could see a puff of dust as the two collided. Suddenly, everything was alright. I knew why I was here. Life was grand. Exhausted, we picked up our bikes and started them. We rolled another smoke and then swung our legs over our iron horses. We put the hammer down to make it home just as the sun disappeared.

11
Normanton

After about a month in the bush, the boss gave us a weekend off, the first time we had a weekend off since I had arrived. A few hours away in the closest decent-sized town, they were having a rodeo, and we were all headed there. I jumped into an available ute seat, and we were off. This was the first time I was exploring Australia not from a horse or motorbike. I was really getting to see the countryside.

When we arrived in the little town called Normanton, it seemed to bustle with people everywhere for the rodeo. I likely felt this way simply because I had been living in a tiny Aboriginal outpost for some time now. We stopped at the Purple Pub. It was my first time sitting in a pub and listening to some true Australian music, a singer named Slim Dusty. This was also the first time I had drank a beer legally in Australia outside of the prison-like pub back in Kowanyama.

With a convoy of vehicles, we found our way to the arena where rodeos would happen for the next two or three days. It was nothing new to us to just find a flat piece of ground and sleep there. We had been living in the bush for so long that it seemed normal. We had just transported most of our rough little stock camp straight into town from the bush.

We set it up in the middle of an old horse racetrack where other people were camping. No doubt we all looked rough. Even after attempting to take a nice, warm shower and clean up, it was quite apparent when we showed up that we had come from deep in the bush. We were ringers from the top end.

I helped some of the young station hands I had met prepare for the bull riding. It was their first time riding, and they were quite nervous and needed a little help getting their gear set up. I knew a bit about this bull riding stuff. I rode bulls most of my life, eventually competing for three years at the collegiate level.

To help them, I climbed on the back of one of their bulls to adjust the size of their bull rope. The bull somehow reached up with a back foot and smashed my foot against the side of the chute. I paid little attention to it because I had been drinking for a few hours. In the bush, we did not wear shoes often. Since we were in town, we all wore shoes. I was wearing a traditional pair of Australian shoes called "Volleys." These shoes are almost paper-thin tennis shoes that offer little protection. After that bull smashed my toe that day, it stayed purple for nearly two months, all the way until I returned to the United States.

Both men I helped did well in the rodeo. After the rodeo ended, the shenanigans really started. It was getting dark, and music was booming out of people's camps. The alcohol was really flowing now. I was rolling myself a new cigarette every five minutes, it seemed. We were letting our hair down. That night there was a rodeo dance, a live band playing country music where people would be dancing.

I grew up around rodeo and honky-tonk bars. I learned to swing dance fairly well. Everyone from camp was waiting to see the dancing skills. During this time, I met an Australian girl who had recently spent some time in the United States. She was working on one of the cattle stations that our company had been contracted to help muster. I had seen her before but never talked to her. When the information got out that she also knew how to swing dance due to her time in the United States, it was a date with destiny.

When the dance finally started, I walked into the rodeo dance with Frost, the man I had tripped over in the dark while being eaten by ants. By Aboriginal standards, he was a big man and had been an incredible boxer in his younger days. He earned his nickname "Frost" by icing people, knocking them out. We strolled into the pub together, looking around for people we knew. Frost quickly started introducing me to his "cousin brothers." Cousin brother meant they had some sort of relation between them, sometimes very distant, sometimes actual cousins, and sometimes I think the term just meant someone they had known forever.

Frost was catching up with family, so I wandered around alone, looking for more of my mates. I rarely paid for a drink in many parts of Australia. If there ever was another American who had been in Cape York, nobody had ever heard of them. People were still incredibly shocked to hear my accent in this rough, dusty little town in the middle of nowhere. Strangers often offered to "shout" or buy the next round.

Finally, I found my group of friends in the raging and crowded pub. They had been waiting for me. The second I walked up to them, one of them grabbed my drink in my hand. Another shoved me out onto the dance floor. I ran directly into the Australian girl who knew how to swing dance.

Swing dancing is a quick, twirling dance, usually between a man and a woman. It is the most popular dance in the western

United States, where I grew up. However, after traveling to many countries, I realized that other countries have a remarkably similar type of dance. These dances just have different names and distinct styles.

Once I had found my dancing partner, I spent much of the night on the dance floor. Some Australians know this kind of dancing, especially at rodeo dances. We must have been tipsy enough to look like a mess, or we really put on a show. We were given a large part of the hardwood dance floor as soon as we started dancing. Spinning around the dance floor, I could not help but notice the large number of eyes that seemed to be fixed on us. When we finished our first dance, we returned to high fives and multiple people buying us drinks.

It was the beginning of the end for me. After that, the night turned fuzzy, and I remember only parts of it. But, I remember it was one of the best nights of my time Down Under. As a crew, we finally got to let loose after working in incredibly tough conditions for a long time. Everyone was smiling. Everyone was falling in love. It was a night I do not need to remember to know the important things that happened.

The next morning, I woke up in my bedroll, the sun was barely up, and it took me quite some time to orient myself. I had been waking up far before the sun for so long that it gave me a fright to see the sun was up. Even more shocking, I was not alone in my bedroll. As B.J. Barham said, we were "tangled up like Christmas lights." I peeked my head out to see what was going on, and the crew was standing not far away around a rolling fire.

The sun was creeping into sight. It was still cold, and everyone was smoking cigarettes and standing around the fire. I rolled my first cigarette and used shaky hands to bring it to my mouth to light it. Someone shoved a bottle in my face. I was told we had to get the ball rolling again. It was only seven in the morning, and everyone had taken pulls off the bottle.

The following night got even more out of hand than the first one. There was no rodeo dance this night. Instead, everyone loaded up and headed into town. It was far enough away that we certainly needed to drive. We got to a little pub where everyone had gathered. People recognized me and wanted to talk to me. I began turning down free drinks because I could not handle them anymore. I remember begging a pretty blonde girl to dance with me over and over, and she refused because I was quite drunk, and she had a boyfriend.

After that, the next thing I remember was waking up in the grass behind the pub with no idea how I had gotten there. I do not remember if I was hit or dragged or just laid there myself, but I was still quite drunk when I woke. It was still dark, and the pub was still full of people, so I did not know how long I had laid there. I walked back inside when I ran into our bull truck driver, "Mad Rich," and he was following a couple of girls.

When Rich spotted me, he told me, "No use in going back in there, everyone we know left, follow me. We're going to a house party." In my drunken stupor, that sounded fun, so I followed him. There was no party when we arrived. Mad Rich disappeared with one girl, and I realized I did not know how to get home.

Then the drunkenness took a back seat to fear. I had wandered behind him, not paying attention to where we had gone. I also had paid little attention to how we even got to the pub from the rodeo grounds. I ran. I ran roughly in the direction I thought we had come from. When I finally reached the edge of town, there was one main highway heading out of town. After asking a drunk man sitting on the corner, he told me to walk down that highway to the rodeo grounds.

I saw no lights in that direction. I saw nothing but pitch black. It was the only information I had, so I started walking. I then realized some guys were following me. I could not hear exactly what they were saying, but I could catch a few words here and there. After the third time hearing them say, "This white fella," I decided it was time to run. I ran down this pitch-black highway in the middle of the night with no phone or communication with the world. I was wondering if this was how I would die.

Outside of town, I had to pass through a graveyard that was on both sides of the road. I had stopped running, almost out of sight of the lights from town. I got goosebumps walking through the graves. I was terrified, and then I heard a stick snap. I looked over quickly just to see a kangaroo had done it. Then another snap, another kangaroo, and it kept happening. Lots of funny noises were now coming from the dark. I realized either every kangaroo in Australia was there with me, or I needed to get the hell out of there. So,I ran again down the centerline of the dark highway.

A figure appeared on the road ahead of me out of the darkness. I kept walking toward it until I was very close, and I could see it was an old Aboriginal man. He was not walking anywhere, just standing there wearing only an old dirty pair of shorts. He had long, wild, black-and-white hair and a big beard. He greeted me like it did not surprise him at all. At this point, I was quite skeptical that I was headed in the right direction. I thought I would sleep in the ditch and find my way in the morning.

When I explained my situation to the old man, he just smiled. He said nothing and pointed down the road, signaling that I was headed the right way. I thought little of this at the time. The rest of the time I was in Australia and told this story, people questioned if that man was real. Most Aboriginal people would tell me it was a spirit guiding me home. Others said I was just drunk. Regardless, I still get goosebumps thinking of that man.

After I ran some more, I finally reached the rodeo grounds. I crawled in my swag and instantly fell asleep. The next morning, in the daylight, the distances were not anywhere near as far as I remembered that night before. The following day, we packed up all our things. We realized we had lost my mate, Frost. Frost left his shoes, swag, and everything else behind, but we could not find him.

After a two-day bender, I did not feel so good. I did not feel like sitting in the middle seat of the ute the whole way back to bull camp. So, instead of getting in the cab, I crawled in the back on the flatbed. I laid on top of all the swags we had rolled up and thrown back there, along with Archie's dog. I lay in the warm Australian sunshine as we flew back down the dusty road to bull camp. I fell asleep quickly, only to be awoken a few times by hitting massive holes in the road. The bumps sent me and all our gear airborne, but it all landed back in place.

A few weeks after, we located Frost and sent Archie to fetch him and bring him back to bull camp. Archie picked him up in the same ute I had been riding in the back of. They drove the same road that we had driven a few weeks earlier. While driving, Archie lost control and rolled that Toyota multiple times, injuring both men in the vehicle. Had that happened when I was sleeping in the back a few weeks earlier, I would have died. Because of this accident, Frost thought this was the spirits telling him not to go back to bull camp. Frost had broken his hand in the crash. I never saw him again.

12
Pard

For the most part, Australians have a lot of respect for the American cowboy. While they both have similar skill sets, they often do things differently. One thing that remains the same across these two groups is the love and admiration for a good horse. My boss in Australia was a splendid horseman. For the first time in my life, someone taught me a little bit about how to teach a horse.

Until this point, horses had always been something to me that I got on and rode to get a job done and then put away. The absolute best horses were enough fun to ride that you would occasionally take them out just for fun. In Australia, we rode to get things done. Many of our horses were barely broke. They were not always enjoyable to ride. Sometimes I would have preferred to get off and walk rather than ride these young wild horses.

They gave me a young gray horse that I was told had only been ridden once or twice. Because I was an American cowboy, the crew believed I must be a horse whisperer. The first morning they gave me this horse, it took nearly an hour and the entire crew to catch him in a pen. He was wild.

Getting a saddle on the horse took four men. They did not give me a chance to try him because we had a full day of work to do, so we loaded him straight into the truck. He was half-wild and had no desire to get on the truck. We tied my stirrups above my saddle and ran him through the chute to load him the same way we load wild cattle.

Once we got to our kicking-off point, we had to unload our horses quickly because my issues had already cost us a lot of time. Two young Aboriginal guys were also given young horses. We used tree branches to flap at the horses to make them jump off the back of the cattle truck. Once unloaded, it was time to get on. The only person with a tame horse on the entire crew was a young lady that was waiting on us. We all swung up into the saddle on our fresh colts with pure terror running through our veins.

For a few seconds, everything was ok. The horses stood and let us get situated. We got our feet in the stirrups, and then it was like the popcorn finally hit the correct temperature to pop. Horses and men started flying in every direction. It was every man for himself. These horses hopped, hogged, jumped, and twisted all over the clearing we sat in. Somehow, all three of us stayed on top of our mounts. When it was all finished, we looked up at each other and gave smiles full of admiration. Slightly terrified and incredibly proud, we had weathered the storm.

Throughout the day, we had many more similar incidents. Only one of the Aboriginal men named Smiley fell off. The poor fella fell off repeatedly as the day went on. I believe they called him Smiley because he had big teeth and was always wearing a dirty smile. On this day, his smile had disappeared. Every time he fell off, he needed help to catch his horse. It scared his horse each time he hit the ground, and he could not hold the reins. Everyone else was barely staying on their horses that day. This meant that men on four-wheelers and dirt bikes had to chase his horse until they caught it.

My horse feared everything. Many times, he almost threw me off in a panic. The shadow of the choppers overhead would scare him so badly that he would start bucking. It was everything I could do to stay on him. Staying on top of him was even more difficult because I had grown up riding in a western saddle. I was now riding in an Australian saddle, which is considerably lighter and offers much less to hold on to.

By the end of that day, my horse was so tired that I could barely get him to trot for over five seconds. I was equally tired because I had spent most of the day wondering when he would fling me down. I had also spent plenty of time wondering what would happen if I found myself on that hard, dusty ground. If I could still move, I would have to climb back on this bugger and try it again. So, I gave it everything I had to stay aboard my pony. When yarding up the cattle that day, the cattle went in easily. It had been a long hot day, and everyone was relieved when it was over.

When I released my horse that day, I was so proud that I was really at a loss for words. Green horses, meaning horses with little training, always had scared me. I always thought that there were way too many good ones in the world to be climbing on ones that want to throw you off. But that day, I came to a new level of understanding with riding an inexperienced horse; I could sympathize with those folks who enjoy climbing on them. That day I decided to call that horse "Pard," slang for partner. I told no one that but him. I figured me and him were pards because we were both probably equally scared. Neither of us had much of an idea of what we were doing in this wild place.

That day when I released Pard, I whispered to him, "You take care of me, and I'll take care of you," and then slipped his halter off. My boss had taught me something I will never forget. He would say, "You should always leave stock the way you would like to come back and find them." We never left cattle running around searching for their babies or water. They had everything they needed. We left them settled down, and then we turned our backs to the content cattle and rode away. I adopted this mindset with my horse as well. I never left him in confusion or disarray. I only released him facing me, standing calm without a care in the world.

The following morning, I had to catch him again. This time it only took half as long, and that day he only bucked half as much. By the end of the week, I could walk up to him in the dark most mornings and catch him without him flinching. It was quite rare to catch a compliment on this outfit. This was one of the few times I got one. "You might make a horse out of him yet," someone said. That was a good enough compliment for both him and me. They assigned me that horse for much of my remaining time there. My boss taught me how to make him move using pressure from my feet. I learned different tricks to make him more responsive to me and less responsive to outside monsters, like shadows or puddles.

We certainly still had our ups and downs in those few months. He occasionally tried to throw me off, but I had his number now. He could not throw me. The last time I rode that horse was a terribly sad day. He had become one of the best horses in the string. When I was given my choice to ride any horse in the herd, I began picking him. It is one of the things I am most proud of in my life, the first horse I ever "made." I probably did a lot of things wrong; he certainly had a lot left to learn, but for a few months, we grew together. We bonded. Without the other, neither one of us was much good.

13
Big Headache

I became quite trusting of my horse Pard as the season went on. He was still scared of many things and would still sometimes shoot sideways without warning. Since we rode so much, we had begun to think alike, so I could usually spot something he would not like. I tried avoiding scary things, or at least prepare him for them. There was one instance when I nearly got thrown off. It was a stark reminder that while we were making progress, he was nowhere near a finished horse.

One of our heifers, just a small yearling, had gotten its leg stuck in a fence. I watched it get its leg stuck, so I rushed over on my horse to get its leg out before it broke it. Amid all the chaos, they released the rest of the cattle from the yard. All my coworkers left in a hurry, trying to contain them. So, they left me with this upset animal weighing five times my weight, with a leg stuck in the fence. My horse was terrified.

As I tried to work the animal free, my horse pulled away from me, making the task almost impossible. After five to ten minutes of being drug around by a horse and wrestling with an increasingly aggravated bovine, I finally got the animal free. I was so exhausted, elated, and frankly mentally overwhelmed by the stress of it all. I had not thought about what would happen when the animal got released.

The white Brahman's eyes flashed white with rage. She left me begging for the air to refill my chest, lying in the hot Australian sand. Amazingly, I had not let go of my horse when the dust settled. Pard was calmly standing over me, reminding me how stupid I was. I let my anger get the best of me at this moment. The heifer had left the corrals. She was chasing down the thousand other cattle that had left minutes before.

Over the radio, my boss began questioning me. That progressed to yelling at me to get the lead out of my boots and get back with the mob of cattle. I normally climbed on my horse slowly and carefully. I did this to avoid scaring my horse before I was aboard and ready to hold on. In my blind fury, I stepped quick and hard into the saddle. I accidentally raked my spur across my horse's butt. What ensued was a sight that only my eyes witnessed, but it must have been a spectacle.

He immediately jumped forward, then stopped, throwing me onto his neck, before hopping again. As he threw his head while jumping and hopping, I somehow landed back into the saddle. By some miracle, I stayed aboard. We were now galloping after the mob in front of us at full speed. I had no control. As we blindly galloped through the thick dust, I got my feet in the stirrups.

As the herd came into sight, he slowed. By the time we had gotten close, he had slowed to a trot. Finally, he walked right up behind the cows and fell in exactly where we belonged. Someone turned around and said something like, "Good hustle getting back up here, the boss was getting upset you were taking so long."

They would have found it hilarious if they could have seen the ordeal I just went through. I smiled and said nothing, patting my horse on the neck and telling him, "You take care of me. I'll take care of you." After that, I thought, my horse would not be able to not throw me off. He could get me slightly out of place, but I thought I would get myself back before he could get rid of me.

Weeks later, the crew and I were pushing a large mob of cattle through a small gate, a process that is slow and time-consuming. As some bulls in the group got restless waiting their turn, a few decided they would run in the other direction, directly at me. When something like this happened on our crew, the boss expected us to turn them back into the mob or die trying. I ran after them on my horse. We jumped a dead log as we got close to the three bulls.

Upon the hard landing, a piece of my saddle snapped, and my water bottle came loose. The bottle bounced up toward Pard's head, falling at his feet, inciting a state of pure terror. Then, at a dead run, he jumped sideways, tipping me off to the side with only a small piece of mane in my grip. We looked each other eye to eye at that moment. Then he let loose a few sharp bucks, sending me headfirst into the crusty ground, directly in the path of the three renegade bulls.

I do not remember what happened for the next half a minute. The first thing I remember is my boss sternly asking what I wanted to do. He was not pleased I had fallen off. I went to grab the reins of my horse to get back on. When he saw me visibly stumbling and my hands shaking, he said, "Mate, you're shaking like a dog shitting razor blades. You're not getting back on this horse." He removed my saddle, put the horse into the herd of cattle and drove him home in the mob like a cow.

Quickly, I noticed I could not see out of my right eye, a feeling I knew all too well from my days as a rodeo athlete. I had a concussion. When this happened, the others realized this may be a little more serious than just dusting my britches off and getting over it. The boss radioed one of the chopper pilots, telling him to land and pick me up. I had always wanted to ride in the choppers, but not in this way, en route to the hospital.

I ducked down and ran under the spinning rotors of the chopper. I climbed in and put my cowboy hat between my knees, slipped on the headphones, and we were off. The pilot asked if I was ok. I knew I had a concussion. Seeing that I was already looking better, he took me on a brief detour to spot some crocs from the air. Then, he took me back to base camp where the cook was waiting to take care of me.

The cook loaded me into her vehicle, and we sped into the local village to find a doctor. I still had a massive headache, but my vision had returned. After only a few minutes of sitting in the air conditioning, I was feeling much better. The cook still drove quite fast down the rough back roads, so fast that when she hit one large bump, she hit it so hard that it threw me up into the roof, hitting my head again. She felt awful, but I could not help but just laugh.

From the time I hit my head to the time I got to a doctor, roughly two hours had passed. The doctor who treated me was working for the Royal Flying Doctor Service. They were assigned to this very rural Aboriginal town. She said if I had been slightly more injured, I would have had to go on a plane into Cairns. Cairns was another two hours away by plane.

It occurred to me just how serious even minor injuries can be that far out. It also made sense why when I asked about how to take care of snake bites, they simply told me to "lay down and die peacefully." There was no way to get to a hospital in under four hours. Luckily, as I suspected, it was a concussion and nothing more. The doctor gave me some pain pills and told me no riding horses for a few weeks and no taking part in any other activity that risked me hitting my head.

When I returned to camp, I told the boss what the doctor had said. He smiled and said, "No, you earn your keep around here. You will be on that same horse before sunrise tomorrow. You will teach him not to do that again." It was exactly what I expected him to say. So the following morning, I did just as I was told. Pard and I had no troubles the following day or any other day until I left Australia.

14
That's a Ringer

Note: In Aboriginal culture, people do not speak someone's name after they have died. This man's name has been removed from the book out of respect for the dead.

While at the Normanton rodeo, the crew had recruited some additional help for our bull-catching crew. Three new men arrived shortly after we returned from the rodeo. Two of them were guys from the south of Australia. They talked about how they had been best friends from a young age. I lived with a crew of rough men, some of the toughest I have ever known. When these two fellas arrived, they instantly had everyone's respect. I did not understand why this was. It was something you saw little of in these parts. I knew they must be legends, and I took everyone's word for it.

The third recruit was an Aboriginal man. What he lacked in size, he made up in character and heart. He was one of the funniest men I have ever met in my life, always wearing a smile on his face, hollering and cracking jokes, keeping everyone in a good mood. In the coming months, all three men would make impacts on me that would last a lifetime, things I think back on often. Lessons of the sheer power of will, grit, toughness, heart, and just how far those things can take you in this world. I learned the respect these qualities can earn you.

The story was that someone offered the Aboriginal man this job while he was incredibly drunk. He had told the boss he could ride anything with hair on it. The next morning, he awoke in the back of a vehicle heading down a dirt road with two white men in front. It shocked him to see this. They had to tell him the entire story of what had happened the night before, for he could not remember it. When they had finished telling him the story, he sat back for a moment before screaming, "BY GOD, I GOT A JOB, WOOHOO!" and that was how he ended up in the middle of nowhere, working with us.

The Aboriginal man was a horseman. He told stories of learning about horses from a young age. He had learned from his elders and family members. He was given a half-broke horse like the rest of the low men on the totem pole, including me. Because of his confidence, he was given possibly the most rotten of all the horses we had available, and he loved it.

The other Aboriginal men and I were given young horses that needed time and education. He was given a mare that was a bit older. She already knew a lot of things, but all the things she knew were bad habits. She knew how to buck, throw a fit, bite, and all the other terrible things spoiled horses know. I do not remember where exactly this mare had come from. The mare was different-looking. She was not like the rugged, small horses most of us rode. She was tall, fat, and sassy.

One day we had to trail about one thousand head through a massive dry riverbed. The riverbed was lined with old, dead trees, thickets of dense brush full of long, sharp thorns, and green ant nests. It was awful getting the cattle through these short, sharp trees. We crossed the rocky, uneven, riverbed and went back up the other side through the same mess of pointy trees. When we reached the other side, we stopped the cattle in a big clearing to let them rest. We pulled our lunches out of our saddlebags and devoured our food.

The sun was high in the sky by this time. It was hot. Almost everyone had battle wounds from busting through the prison wire-like barrier that lined the riverbed. We had no option other than to hug our horse's necks and let them find their way through the brush. Cuts across the arms, legs, and face burned with salty sweat. The Aboriginal man always packed just a few oranges for lunch and nothing more. He never carried water either. Instead of pulling his lunch out, he had other plans. His horse had been giving him trouble all morning. He was going to straighten her out.

Once mounted, he pulled the horse's head from side to side to flex and loosen the horse's neck, making her easier to control. He then asked the horse to move forward by touching her with the heels of his boots. The horse did not move. Now he bumped his heels into the horse, but nothing still. Finally, with the horse refusing to budge, he straightened his legs and kicked with both feet.

This time, the horse moved. She moved up and down and side to side, trying everything she had to throw the man off, and it worked. He hit the ground with a THUD and groaned but got back up. The man, without hesitation, climbed back on the sour old mare, and round two ensued. She had learned she could throw him off, and now she was going to do it again and again and again.

The second or third time he was thrown, it became personal for the man. It was no longer about teaching her, no longer about getting the cattle to the yard that day. It was no longer about looking tough in front of the crew. It was a dogfight. Aboriginal men in this part of Australia are famous for fighting. They have some of the most gruesome, bare-knuckle street fights I have ever seen. Their trademark is that neither fighter ever backs down.

The mare that picked a fight with this man did not know what she had gotten herself into. The second or third time she had thrown him, the man had taken off his dusty, torn-up shirt. He then took off his belt and wrapped it around his hand. At no point during the ordeal did he ever take a cheap shot at the horse by striking her from the ground. He calmly climbed back up into the saddle each time, where the battle started and would end.

While most men would have called it quits or at least asked for help, he faced his battle alone. As the mare jumped in every direction, giving him all she had, he took his belt and smacked her on the butt each time she hogged and jumped in the air. He was letting her know he was there to stay. When the dust settled after lunch, He had won. He sat atop his bronc with pride.

Most of the crew looked on in awe at the display of heart we had just seen. No doubt there would have been a round of applause and cheers if it would not have startled our wild cattle into a stampede. The boss was a man of little praise. It was rare to hear him compliment anyone. He rode past me that day and said, "If you want to see what a ringer is, that is one right there," as he pointed at the thin Aboriginal man.

It gave me chills when he said that and still does to this day. It was like watching a moment that takes you back in time. This was what legends are made of. Throughout history, most people avoided conflict. But somehow, the past seems full of heroes, legends, and hard men. We tell stories about these men and someday hope to resemble them.

People do not tell stories of ordinary men; they do not tell stories of cowards. People tell stories of legends, and I witnessed the beginning of a legend. This man was thrown off probably seven or eight times that day, each time landing painfully on the hard ground. Yet, he never lost his cool, he never lost his pride, and he never lost his will to finish something he started. Legendary.

15
Town Dogs

Most of the men I worked with had vehicles, and occasionally they would drive into town for something. If there was an open spot in the truck, I would always ask to jump in and go with them. In town, there was only one supermarket. It was also the only place to buy fuel. This place had fresh fruits and vegetables occasionally. Those were the only fresh things in that store.

Almost everything else in the store was canned or processed, so it had an exceptionally long shelf life. Nearly all the soda, potato chips, and snacks were far past the expiration date. The only way the store could stay stocked was to buy what stores in the city discarded. Most of these things were over a year past date, but everyone bought them regularly, so I did as well.

It was a special treat to go into this place to buy something for ourselves. Out at camp, we did not get sweet things or cold things, little things we take for granted. Items that are readily available in most built-up places became treasures to us. We would get candy bars, chips, cold drinks, and ice cream, all far past what the manufacturers deemed appropriate to eat them, but we felt like kings.

The entire time I was in Australia, I had no cell phone. I had not known how to get an Australian SIM card for my phone. This meant no contact with the outside world unless someone would create a hotspot on their phone. The hotspot would allow me to access the internet. I could talk to my friends and family through email or social media. I did this every chance I got, but that was usually once a month if even that. Eventually, my phone stopped working altogether. So, I occasionally borrowed other people's phones for twenty minutes.

One afternoon, two men from the camp headed to town and invited me to go with them. They were hoping to find some weed. Some of the men I worked with enjoyed the devil's lettuce. In Australia, they called it "hoota." They said if I went, I could borrow one of their phones while they conducted their business. So, I rode in the middle of the ute all the way to town. When we finally reached cell service, they handed me a phone. I began typing things to people as quickly as I could. Unfortunately, in this remote part of Australia, there were few options for places to buy weed. So, we looked for a man they had heard of named "Bulldog."

It is hard to paint a picture of just how sketchy this was. Three white men, the only white men in the whole town, driving around at night asking total strangers where the local drug dealer's house was. We drove around for almost an hour before they gave up. Just then, we passed a man. They thought to give it one last shot and ask him where to go.

He knew and gave us directions, and we tried to follow them but could not. Finally, my coworkers asked if I would get out of the ute and sit on the street corner, where I could check what I needed on the phone. Unfortunately, the ute only had room for three people, so I had to get out for them to pick up this stranger so he could show them where to go.

I got out and sat under a streetlight on the edge of town. It was late, completely dark at this point. I was now all alone under this streetlight that provided little light. As I sat there nervously reading emails from friends back home, I noticed a stray dog wandering closer to me. This town was full of stray dogs, many of them very unhealthy. The dogs had mange or three legs, were blind, or had at least one major issue. Many of them had huge, saggy teats, showing they had birthed many litters of pups. Most of these dogs had never been around a white person before. While some of them paid no attention, many dogs would begin barking and growling the second they saw a white person.

As the stray dog got closer, I got a little worried. Not that it may bite me, but that it might cause a commotion and draw attention to my already sketchy situation. Finally, it had gotten so close that I felt I needed to make sure it knew I was there. If it came closer and I startled it, it surely would bark. In a low, almost whisper, I let the words "go away" slip out of my mouth, and pandemonium ensued.

I scared the dog half to death and when it saw I was white, it lost its mind. It began barking like crazy. I was afraid of it drawing people's attention to me. But, what the barking did was much worse. It sounded the alarm. It seemed to raise every dog in the town, and they came running. There were big dogs, little dogs, fat dogs, skinny dogs, dogs missing legs, dogs missing eyes, black dogs, white dogs, and red dogs.

Dogs just seemed to keep coming out of the shadows. Instantly, I was surrounded. I knew I should not run or make sudden movements because this would excite them. When dogs get into a group like this, they do not act as they would normally act alone. In a group, they regain that pack mentality. Groups of dogs become bolder and more aggressive, feeding off each other's energy.

I looked around, hoping to see that ute in the distance coming to get me, but they were nowhere in sight. I realized my only option was to climb the light pole I was standing under. It was a big smooth pole with nothing to grab onto. I bear-hugged the pole and climbed, hoping I could hold myself until my mates returned. I could barely climb the pole and knew I could not stay there long. I remember thinking I was going to die being eaten by street dogs, all because I wanted to check Facebook and my buddies needed a bag of weed. There were so many deadly things here to kill me, and it would be the sick and crippled street dogs in this wild, northern part of Australia that finally got me.

Just then, I saw headlights. It was my mates. I had never been so relieved. When they pulled up, they were rolling with laughter and asked how I was doing. I said nothing. The terrified look on my face told them all they needed to know. They said they were not getting out and getting bitten themselves. They said I better make a break for it and jump on the back. I gathered my thoughts, took a deep breath, and bolted.

One of the biggest dogs lunged as I moved. With a closed fist, I swung with all I had and hit him in the nose. I then began hurtling weenie dogs and swatting dogs off that attacked from behind and ran for all I was worth. As I got close to the vehicle, I started deciding where I was going to dive on. They played the meanest trick on me anyone had played so far in Australia. They pulled forward just fast enough that I could not reach the back of the truck. I had to sprint to catch up to them and fling myself on.

16. Many Ways to Remove a Man From a Horse

The crew I worked with were masters of many things. One art form they particularly enjoyed was making someone fall off their horse in different ways. Everyone loved a bit of rodeo action at any given time during the day. If the crew could do anything to promote a horse trying to throw a rider on the ground, they would. But, of course, the chopper pilots had the greatest opportunities.

The pilots were the eye in the sky. There was no escaping them if they wanted to mess with you. The men on four-wheelers also could scare someone's horse easily. They could send your horse into a wild panic, bucking across the flat and through the trees. All the horsemen could do was hold on for dear life.

One trick that chopper pilots would employ against ground dwellers was using their shadow. Horses and cattle do not actually look at the helicopter moving above them. It is mostly the loud sound that causes them to move. However, the animals can see the shadow of the chopper on the ground. So, given the right sunlight angle, the chopper pilots can use their shadow to their advantage.

This phenomenon can also be seen in Wyoming, where I come from. Many ranches have massive wind turbines that spin high above as cowboys work. These do not bother cattle or horses, but the shadow they create can. I have heard stories of men riding young horses through the shadow of the spinning blades above them, not paying attention. The shadows of the blades scare their horses, getting them flung off. Chopper pilots did this whenever the time was right. The person on the ground would not see it coming at all. Before they knew it, their horse was bucking, often throwing them off like a lawn dart.

Another common way to get removed from your mount in a hurry was on behalf of the men on four-wheelers. These four-wheelers have massive bumpers built on the front so they can smash into almost anything. When riding through a patch of thin young trees, riders must always watch for men trying to play tricks on them. The four-wheelers had power, weight, and protection. They could easily knock down trees the width of a broom handle or even larger. If hit at the proper angle, they could knock the tree down directly on someone before they even noticed what was happening.

The tall, thin trees offered the driver an impressive attack range when they felt the need. Even if the tree narrowly missed the horseman, the slapping of the leaves into other trees and the ground made enough noise to scare most horses. Only a few trees hit me before I learned my lesson. I always watched in every direction when riding through trees. Each time they hit me, it would send my young horse into a bucking panic. We would bounce off trees and crash through others until he had calmed down.

After getting bucked off once and going to the hospital in a helicopter, I now understood the severity of even minor injuries. I could see just how important it was to stay aboard my horse and avoid injury. This made me a much better rider, because I got to learn much like the old cowboys who brought cattle up the trail from Texas. They climbed aboard the rough ones and took pride in this. However, they also knew just how severe a bad horse wreck could be in those times. They rode through territories without doctors or aid. It could surely mean death.

Teddy Blue Abbott spoke of the old-time cowboys in his book, We Pointed Them North. "They used to brag they could go anyplace a cow could and stand anything a horse could," said Abbott. This also made me understand just how far "jokes" would be taken by these men to have a good laugh. I believe this mentality made them such hard, tough men, hence, their disregard for safety and the rules just to have some fun. Yet, it somehow almost always ended up being okay. Never, to my knowledge, was anyone ever seriously hurt due to these pranks.

17
Just Jump in Back

Often our work kept us out long after dark. We would then commute back to the camp. The dark brought to life an entirely new bush. Some animals were more active, some were less active, and the dark made it a bit scary. The dark was when many people went pig hunting with flashlights because the pigs were more active. You could often see the eyes of crocodiles reflecting the light back when you shined on the water. Many snakes became more active in hunting at this time of night. This meant being especially careful where you walked and always sleeping off the ground, if possible.

The vehicles that most people used in the bush had massive front bumpers called "roo bars." They often had massive off-road lights to illuminate as much of what was in front of them as possible. The roo bars wrapped around the side of the vehicle coming back to the driver's door, so the fenders were also protected. The kangaroos and wallabies were more active at night. It was common for people to hit them at night; thus, the name roo bars, protection from kangaroos.

Most of the Toyota trucks only seated three people in the front, so I almost always was made to ride in the back. I rarely minded. The fresh air felt good, and I could stretch my legs after a long day. One day we finished branding late, ten or eleven at night. We loaded our gear and headed home. I jumped in the back.

The back was full of all our branding equipment. Huge drums were full of diesel and what we called "av gas," or fuel for the choppers. We also had a branding chute, irons, propane bottles, a stove on which to cook our lunch, and some plastic gas cans. With all the supplies, I had little room to sit.

I squeezed between the fuel drums and sat on a gasoline can. It was not comfortable, but it was the best I could find. I sat in the center, which would hopefully keep me from bouncing off the side. The boss drove. He loved to drive like a bat out of hell, whether he was in a hurry or not. Like always, the guys up front decided it would be funny to play a trick on the American stuck in the back in the dark. So when they spotted a dip in the road, the boss sped up and aimed directly at it. Meanwhile, in the back, I had no idea. I never saw it coming. I faced backward the whole time to avoid the wind in my face and eyes.

He hit the dip hard. He later told me the dip was much bigger than he had thought, and the headlights had tricked his eyes. The truck went airborne, and along with the truck, so did I. All the sharp and extremely flammable and fragile things in the back levitated with me. Av gas is notorious for blowing up on people and killing or severely burning them. I was sitting next to a drum that was bigger than me. We flew in slow motion. Like something out of a movie, I watched all the heavy pieces of gear float around me in the darkness.

The truck came crashing down on all four wheels. The boss immediately came to a stop. He scared himself and everyone in the cab with the bigger-than-expected bump. It had thrown me up and then dropped me so hard that it had momentarily knocked the wind out of me. When someone in the front rolled down their window and hollered, "You alright back there, Texas?" I was still trying to find my air and could not answer. They had a terrible feeling of dread come over them. They thought they had bucked me off.

Finally, I gasped and said, "Yeah, all good," and they breathed a sigh of relief. Now I was soaked, and I was not sure what was leaking in the bed. Every fluid back there was flammable. So, I took out a light and searched around to find out what was leaking. Luckily, it was the least volatile of them all. It was diesel, the best-case scenario. I had landed on the plastic gas can, breaking it with my weight. A large pipe had hit my ribs, leaving a bruise that lasted for weeks.

The pranks like this were usually lighthearted in intention. However, if something had gone wrong, the consequences could have been deadly. We finished driving home with the truck dripping from the back. I sat smack in the middle of it all. No harm was done besides a broken fuel drum, cracked gas can, and some bruised ribs.

18
Always Carry a Spare

We needed to cover tens of thousands of acres as we looked for cattle to capture. This often meant dividing and conquering. It was the best way to cover as much area as possible, but this also meant being alone a lot. So, you had to be able to fend for yourself and make intelligent decisions. It could be life or death when alone in this remote wilderness. Cell phones did not work here unless you had the expensive satellite phone.

Even satellite phones only worked when you pointed the antennae straight north toward Asia. This meant our primary source of communication was with radios, "two ways," they called them. The chopper pilots had them, and most people had them in their vehicles. We carried handheld radios to easily hear crewmates and respond quickly while on a bike or horse.

The problem with these radios is that they were quite expensive. They were high-quality radios. So, we did not have enough for everyone to always have a radio. But if you worked on the buddy system, usually between the two of you, someone had one. I had difficulty understanding the thick northern Australian accent and slang words in-person. Over a crackling radio, while riding full speed on a dirt bike, they might as well have been speaking a different language to me.

We fixed hundreds of kilometers of fence while doing this job. Sometimes that comprised of simply driving along it to ensure there were no broken wires. Other times that meant rebuilding entire stretches of the fence where floods, wild horses, bulls, or father time had torn it down. We loaded our fence repair gear in the back of a bull catcher and set out for a full day's work, usually in a group of two or three men. It was a job that nobody enjoyed. It was not glamourous, but it needed to be done. We usually spent at least one day every week or two doing this.

One day Spook and I were assigned to go repair the fences on a large paddock. This paddock was far from camp. We loaded everything we expected to need in the bull catcher and headed out. Just as we had reached the farthest corner of the paddock, we heard a loud PSH PSH PSH. It was the sound of one of our tires going flat. Realizing it instantly, I yelled to Spook to turn and drive toward home as fast as we could on the failing tire.

We spun around and drove as fast as the ancient diesel Toyota bull catcher would carry us. The tire went flat in less than a minute. We had neglected to check all our gear before we left. After coming to a halt in the hot dust, we looked around for the spare tire, jack, and wrench to put on a new one. We were missing every single one of them. We were screwed.

Spook and I rolled a smoke and sat there wondering what the best thing to do would be. It had taken us over an hour of driving to get to where we were. We were incredibly far from camp or even a road. I asked Spook if he thought his radio would reach camp at this distance. He laughed and said, "I don't have a radio, mate. I thought you had one." Then I realized we were in deep trouble. It took us half an hour to decide if we could drive home. We contemplated driving on three wheels. Finally, we decided there was no other option but to walk. Luckily, we had packed heaps of water, a gallon apiece, that would last us the whole day.

After deciding we had to walk back to camp, I gathered some sticks and configured an arrow next to the buggy, pointed in the direction we intended to walk. We grabbed our water jugs, rolled another smoke, and started walking. We did not follow the road because it curved through the gullies and dry riverbeds. The road made the trip much longer. We just pointed ourselves toward where we thought camp was and walked in a straight line.

Spook was not much older than me, but he was a seasoned ringer. He had been in situations like this before. It was slightly comforting to know that. I grew up in the mountains, where I knew exactly where north was at any given time, and going down in elevation almost always meant finding water and your way to help. In northern Australia, the land is flat as a pancake with some washouts and dry riverbeds scattered throughout. The termite mounds and Juju trees are the tallest things in the entire area. Juju trees are filled with thorns and green ants. It all looked the same to me. Other than the sun, it would be impossible for me to determine north.

Spook told me stories of people getting cooked to death by the sun in rural Australia in situations like this. Because we had water, our situation was drastically improved. We often took smoke breaks in the small pieces of shade we could find to keep from overheating. We walked and told stories of the places we had seen and places we would like to go. We spent the whole day walking, hoping we had not gotten lost or turned around.

When I was young, my father was a wildland firefighter. He was also on the search and rescue team in the Rocky Mountains and was an avid outdoorsman. From a young age, my father drilled simple survival tips into my head. I knew that the best way to stay on course was to pick something as far off in the distance as I could see, then walk to that point. Otherwise, you can end up walking in circles. We did just that the entire day.

Finally, when our water got low, Spook and I thought we could hear the low rumble of an engine in the distance. We knew we must be close to a road. So, with haste, we moved toward the sound of the engine until we found ourselves on a road. Slowly bumping down the road, we could see an old Toyota land cruiser approaching us. We waved it down, and it rolled to a stop, bringing with it a massive cloud of dust. Four Aboriginal men sat inside the old cruiser. They carried with them fishing gear. They were headed to the big river to fish for a few days.

The men looked slightly startled by us. Two white men had stumbled out of the bush covered in dirt from head to toe. We wore dirty tatters of what used to be clothing before hard labor and barbwire had destroyed them. The driver said, "How you going, bros?" We told him of our situation, and he invited us to jump in and gave us a ride closer to camp.

The cruiser looked apocalyptic: someone had smashed all the lights and windows out. It had a dent in every panel. None of the tires or rims were the same, but it ran like a champ. The driver laughed, telling us the stories of how each piece of the vehicle had been broken. He laughed the hardest while telling us about losing the windshield. He said, "My missus caught me playing up on her, so she came out one day and bashed my windscreen out with a T-post. Then stuck it through the back window and left. She come back the next week, though." He had cheated on his wife, something that was quite normal, it seemed, in this community.

All of the men here wore little clothing, just tattered shorts. It was almost like a clash between the present and the past. These men took a huge amount of pride in living off the land. They were outstanding outdoorsmen, hunters, and fishermen. For this multiple-day trip, they had brought nothing more than a lighter and fishing poles. They intended to survive that way, and they did it well. They reminded me of the men you would see in movies or history books that lived off the land. They just had a touch of modern technology they used to their advantage.

We signaled the driver where to drop us off, thanked him for the ride, and wished them luck on their time in the bush. Spook and I walked down the long driveway back to camp, arriving late in the day and telling our boss of our misfortune. An entire day was wasted walking home, and no fence had gotten fixed. The following day, we returned to the catcher with everything we needed to repair the tire. We finished fixing the fence without issue.

19
Stoney

Many of the wildest guys I worked with worked on the bull-catching side of the crew. However, one man who worked on the mustering crew really impacted me. He was the second-in-command, behind only the owner of the entire company. He was a product of the bush. He spent much of his young life in the bush, and now he was raising his family in the bush. The man was short and wide. He was not an athletic-looking specimen, but when the time came, he could outwork anyone on the crew. He was stout, double tough. One might say he was built like a boulder. Men called him "Stoney."

Many of the men I worked with were quite impatient. Screwing up surely meant a chewing out or someone blowing up at me. Stoney was not any different from anyone else. He had no patience for me when I was slow. He often called me "Concrete Boots." He was not a man who easily handed out his respect.

While he would never admit it, he seemed to like yelling at me. At first, this made me dislike and try to avoid him altogether. However, I eventually realized that he liked to yell at me because he wanted to teach me more than the other men. While his teaching method was difficult from the student's perspective, it was effective.

Stoney taught me how to tie a bowline knot, the knots we used to secure bulls to trees. He taught me how to knock down and tie up bulls. He would yell at me daily because I was not tying bulls' feet fast enough with my belt. By the end of my time there, I was nearly as fast as any man on the crew, thanks to him. Stoney taught me how to properly roll cigarettes. He was tired of watching me smoke crumpled-up, crooked cigarettes.

Stoney expected greatness from his dogs. He was even tougher on the dogs than me, expecting perfection from them. He, in turn, had some amazing dogs. I learned a fair amount from watching him train his dogs daily. He taught me about butchering when we killed a cow and prepared to bring it back to camp. I learned a bit about riding rank, half-broke horses from him. One day, an old horse who had not been ridden in some time tried its best to throw me off. He rode alongside me, coaching the entire time and laughing when I stayed aboard.

He was a man of great character as well. He always had a dirty joke to tell. He had a strong accent and used language that only folks from Northern Australia ever used. When we sat around the campfire after a long day of work, he had a loud laugh that brought up the mood. This was when he would pull out a bottle of his rum, pour himself a drink or two, and smoke cigarettes until he went to bed.

Much like the old west cowboys, we had no technology for entertainment, so the time had to be filled with the entertainment we produced ourselves. I heard many stories at night sitting around the campfire. Stories of legends from the past and present. People played songs. We discussed matters of business. It was a place where people were valued and things were used. It was refreshing in a world where people are used and things are valued.

The most remarkable thing about Stoney was that he had a wife and son in the bush. His son was young, only two or three years old. He had taken an old truck for hauling cows and turned it into a living area for his family. His wife worked with us daily; she was quite good with horses. Stoney was exceptionally talented on his four-wheeler. They had a nanny that watched their son while they worked all day.

I developed the utmost respect for them, moving their family in the nomadic way the rest of us moved. So many young, single men were not cut out for this type of work. Moving weekly to a new location in the bush and making a new home next to some pond full of crocs did not suit them. So, when they did it as an entire family, I had nothing but respect for them following their dreams.

20
Rodeo Time

Before leaving the Aboriginal community I had come to love, there was one last major event: the local rodeo. These rodeos had almost no rules and were entirely run by the community. If the champion won any money at all, it was not very much. Winning was simply for bragging rights. They could attain legendary status by making a spectacular ride at this rodeo.

Men talked about special rides for decades after they happened. The old men talked about their rides from twenty years ago. It was not a full rodeo; it was only the rough stock events and a few events for entertainment only. Before the rodeo even happened, they had a huge rodeo dance at the Kowanyama pub, and everyone came down for it.

Our crew was the only white people in the pub, and the pub was full. The whole town came out in their best dress. I was told before the dance that I must "dress nice." In this part of Australia, that meant putting on shoes, jeans, and a shirt with as few holes as possible. Our crew was important to the rodeo because we oversaw catching the animals. They did not hire a professional to bring the bulls and horses for the rodeo. Instead, the community leaders asked us to catch wild ones and bring them in, so we did that.

Because of Aboriginal beliefs, we normally could not catch the wild horses called brumbies. The horses were sacred, a symbol of freedom. They paid us to round up only cattle. Sometimes we accidentally rounded up the wild horses with the cattle, but then we let the horses go. For this rodeo, we had special permission from the elders to bring some horses in to use. All the horses that we had rounded up that were old enough to be ridden went to the rodeo.

Our crew started playing pool in the corner and watching some people dance. Many of them were quite drunk even though the party had just started. There was no DJ. Someone just plugged in their iPod. We listened to worn-out nineties country music and rap. They seemed to have Eminem and George Strait on repeat the whole night. Folks from all over the community came up throughout the night to make small talk about our bull catching and the rodeo coming up.

Some girls started coming up asking to dance. The girls were not shy at all. If they wanted something, they came straight up and told you so. I declined over and over, and so did everyone else on the crew. Finally, a girl in her mid-twenties came up and asked my coworker to dance. I was standing next to him with a pool stick in my hand, and he knew I could swing dance.

He looked her right in the eye and told a boldface lie that I almost believed myself. He said, "You know, I've got a wife at home, and I can't be playing up on my missus but see Texas over there? He has told me how much he wants to dance with you all night. He's American, so he's the best dancer in the place." Her eyes lit up, and I knew I was in big trouble. I declined, but she had already made her mind up.

I was dragged onto the dance floor by this girl and pushed by my mates from behind. The good news was the song was nearly over, so I pretended to not know much. We did a few spins, and when the song ended, I thanked her and tried to walk away. However, she was not about to let that happen. She said she wanted a full dance and would not let me go. Nobody had seen us dance at the end of the last song, but now we drew attention.

The next song came on, and she whipped me around in circles and taught me her style of dancing. Quickly, the dance floor cleared off. I realized everyone had stopped and was watching us. My mates whistled and yelled a few things at me. Then, the place erupted with whistling and cheering and yelling. It was one of the wildest dances of my life, being in the middle of the spotlight with hundreds of people watching.

Then came the rodeo with the brumbies and the bush bulls. It was one of the wildest events I have ever watched. Many of the men simply held on any way they could. Very few of them had official rodeo equipment, which you need to try to follow the normal rules. The pure grit and determination of the men riding were impressive. They would not let go until they absolutely had to, which created some wild rides. One man spun completely around backward on the bull. He kept his hands still stuck in the rope, made the eight-second whistle, and got a score.

The judges were some guys I worked with on the crew who had some rodeo knowledge. They completely guessed the score. In a real rodeo competition, the maximum score is one hundred points. That score has almost never been achieved in the history of rodeo. The scores continued to go up and up with each crazy ride until one man's wild ride topped them all. The announcer screamed into the microphone that a world record had taken place. The man scored a one hundred- and two-point bull ride, and the crowd erupted with excitement.

The men then climbed on the wild brumbies. They used a normal saddle, the kind that they used daily at the station while working. In the other hand, they held a bullwhip. This was an event called "The Station Buck Jump." The rules were simple: riders must stay on the horse and, with their free hand, crack their whip three times in the air.

Riders could jump off after the third crack, and the person who made the most difficult ride would win. The horses had never had a saddle or rider on them. So, when they came out of the chute, they used everything they could to get the rider off.

The last event of the rodeo was the money bull, everyone's favorite event. They asked us to pick our meanest bull and tape a hundred-dollar bill to each horn before releasing him into the arena. Whoever could grab a bill got to keep it. Men came out of the woodworks for this event, some wearing protective gear and some not. They knew that they would be a town legend if they could get that hundred-dollar bill. Our team released the bull. Bold men began running at it from all directions, attempting to get close enough for a try.

The bull wreaked havoc, taking out men and flipping them into the air left and right, many of them needing to be helped out of the arena. This went on for ten minutes or more. Finally, the bull became tired enough that the men could grab the money off both horns. In the end, many people were sore, but nobody was seriously injured, and the bull went home with us and was fine as well.

21
The Australian Champ

During the month and a half I worked in the bush, I became close friends with two men named Chuck and Rhys. They were the men from southern Australia that we hired after the Normanton rodeo, the ones who picked up the Aboriginal man who finally tamed that sassy, old mare. They were two of the roughest, toughest guys I had ever met. We got along well because both had spent some time in the United States and Canada riding in rodeos. They knew where I had come from and how different Cape York was from my home.

We had finished mustering for the day, and I was in the back of the ute with Chuck as we headed home. We stood on the flatbed, holding onto the cab of the ute. It felt amazing to let the air blow across my entire body, drying the sweat and removing some of the dust that had accumulated on me after a long day's work. We would take turns opening each gate as we headed back to base camp. They loved to yell at me for never opening or closing them fast enough.

We talked when the truck was driving slow enough that the wind did not deafen us. At high speed, it is as though the words were simply blown away before reaching the other person's ears. I asked Chuck what his plans were, and he asked what my plans were as well. I needed to be in Brisbane at the end of the month, so I could fly home and finish my degree at the University of Wyoming.

Chuck invited me to tag along as they went on a rodeo run to the south. The rodeo run would last two weeks. They said I could easily grab a ride home with one of the cowboys who would be headed somewhere near Brisbane. During my bull riding career, I went on many road trips. I thought there would be no better way to see Queensland than driving across the entire thing. I told them I would go with them.

I was grateful that I was going to travel with them. It made my life easier, and I had a huge amount of admiration for the two of them. They carried a swagger about them that has stuck with me to this day, something I will never forget. Like brothers, they were always up to no good. They always had each other's back, no matter what. Among battle-tested men, they were highly respected.

There was a particular look in their eyes that always appeared whenever things got scary or dangerous. That look was like what you'd see in the eyes of a cornered dog that was ready to fight back. When flight is no longer an option, fight reflex takes over. No matter the situation, they seemed to believe there was no way they were going to lose. It was quite rare that they lost. They carried this immense amount of confidence about them. However, they were still humble, never looking for a fight, but sure to finish anything that started.

I had finished my commitment to the bull-catching company. Chuck, Rhys, and I packed our things on the back of the Toyota: two dogs, all our camping gear, and a drum full of diesel to help us make it there. It barely fit. We had things tied everyplace possible. For the weeks I had known them, they had spoken little about rodeo. They acted as if they had ridden in a couple of rodeos, nothing more. They were both tough and motivated, attributes I assumed had taken them to minor success in the rodeo arena.

We drove south to our first rodeo in the town of Cloncurry. Chuck told me the story of how he had won the bronc riding in Normanton the month before, where we had first found them. I said, "You won the bronc riding? I thought you were a bull rider." He giggled and replied, "I am. Rhys entered me in the broncs as a joke. I would not turn out like a coward, so I got on and won the whole thing my first time." I thought he was lying until I noticed he had on the champion belt buckle as proof. I had just never noticed it.

When we arrived, Rhys had little time to get ready, so we parked and he ran to get prepared. I helped Rhys pull his saddle tight on the horse they had given him to ride. I watched him use duct tape to tape back on an essential piece of the stirrup. After seeing that, I wondered if he knew what he was doing.

He asked the stock contractor, the horse's owner, how much rein he should give the horse. The contractor replied, "I'd give him X and four. He might even take more." For those who do not understand bronc riding lingo, the owner told Rhys where exactly to grab the rope tied to the horse's head. What he had said indicated that this horse was powerful, and it dropped its head down to the ground when it bucked, making him awfully hard to ride.

When someone rides bucking horses, the rope in their hand is partially a tug of war with the horse. You must grab the rein in the right place or the ride is much more difficult, almost impossible. Rhys looked the owner in the eye and told him, "I think I'll give him X and one, and he can have more if he likes." This was a bold statement. He was claiming he could outmuscle this horse and that the owner was wrong in his prediction.

If the horse took more, he would need to slide the rein through his hand just enough to balance. If he failed to do this, it would surely send him flying high in the air over the front of the bucking horse. He could be badly hurt. Only the most elite bronc riders can let the right amount of rein through their hand when needed.

The comment Rhys had made was a cocky one. It made me wonder even more if he knew what he was doing. As he climbed on the back of the large, gray bucking horse, the announcer told a little bit of Rhys's story. I stood there in awe. "This cowboy is a multiple-time Australian champion bull rider. Champion bronc rider, Australian all-around winner. He has won rodeos internationally in the USA, Canada, and Brazil. He's won go-rounds in Houston and Denver and won the Missoula PRCA rodeo…" The announcer went on. I could not even remember all the things he had won after hearing that.

Rhys had failed to mention any of this in the weeks we had spent side by side. His bedroll was close enough to mine that I could have thrown a shoe at him to wake him up if needed. Some might say he was one of the best rodeo cowboys in Australian history. All we ever talked about was learning how to catch wild bulls and train dogs.

Rhys smiled his famous smile with one broken front tooth. Then, he got a look on his face as if he were going into a fistfight. He nodded his head. They opened the gate, and he came flying out into the arena like a keg of dynamite, BOOM. He never missed a single move this horse threw at him. When the buzzer sounded, he leaped off the horse as if it were nothing. The crowd

him down. Then the next steer came, and I pulled him down by his tail the way they had taught me in the bush.

All the competitors and spectators had taken notice; two drunk guys at the end of the arena were throwing down the steers that the men on horses had missed. One man rode down to us on and said he had a spare horse if we wanted to try it from horseback. I decided I better not since I was drunk and I wasn't wearing shoes. This did not slow Chuck down one bit. He handed me his beer, and I walked and sat down next to some girls with lawn chairs in the arena.

While he was gone, I watched the dog for him. I had taken his short leash and clipped it to the elastic waistband of my shorts. I sat down with my beer, Chuck's beer, and the dog, and watched as Chuck got ready to chase the steer. He nodded his head and came flying out after the steer. It ran right down the fence line, and he jumped right onto its head and twisted it down easily. The competitors watched in awe.

In a drunken hurry, Chuck had nodded without his hazer even in the box. A hazer is another man on a horse that runs along the opposite side of the steer to keep the steer running straight. Chuck had just completed a run, drunk, in shorts, with no shoes on. He had done it without a hazer, and done it faster than many of the men in the competition. After seeing this, the men offered him a second run, but this time with a hazer to see if he could do it even faster.

Chuck exploded out of the starting area. He leaned off the side of his horse as they thundered down the arena at top speed. He grabbed the horns of the steer with all his might. Something was wrong. Suddenly, he was pulled off the horns. Being barefoot, he had put his foot through the stirrup of the saddle and was now stuck. The horse dragged him around the arena at top speed, and everyone ran to his aid.

Since I was sitting in the arena on a lawn chair, I was one of the closest people to him. No matter what it took, I was going to save my mate. I stood up and took off, running toward him. I had totally forgotten about the dog tied to my elastic shorts. He did not even have time to get up by the time I hit the end of the chain. It pulled my shorts down around my ankles, exposing my bare ass for all to see. I tripped and fell face-first onto the arena floor. I still had both beers in my hands, and when I landed, they shook and shot like geysers, covering me in beer.

Chuck's foot eventually slid out of the stirrup before anyone could help him. He came out unscathed. The biggest injury was to my ego, as I had fallen in the middle of the rodeo arena. The task he had accomplished was nevertheless impressive, earning him respect from all the people who had witnessed the spectacle. When he returned, he asked why his beer was empty. I had to explain to him how I had made myself look like a jackass in front of everyone trying to help him, and he cackled.

23
Big Roo

The next day of the Cloncurry rodeo, it was Chuck's turn to ride in the bull riding. He drew a bull that was one of Australia's best bucking bulls named "Chuckles." He, unfortunately, threw Chuck off before the whistle blew, and it was no score for him. It did not matter; everyone was still in good spirits. We wandered over to the bar located on the rodeo grounds.

We found Rhys wearing a massive headdress. He was howling at the moon. A small crew had formed around us. I remembered some names and made a few new friends as the night went on. The more whisky everyone got into their system, the more they wanted to talk with me. Everyone wanted to find out what I was doing in this place.

At first, my response was to tell people, "I'm a ringer from the top end, a professional bull thrower." This was a quite cocky statement to most Australians. There is a song written about this by the famous Slim Dusty. People from the Cape are highly regarded as some of the wildest and most talented stockmen in the game. When I said this to people, they often thought it was a joke. When they discovered who my traveling partners were and others vouched it was true, people were astonished. Nobody had ever heard of an American doing such work in northern Queensland.

Giving people the full truth soon became boring, so we schemed a funnier response. The Australians had given me the nickname "Texas." While I repeatedly explained that I was not from Texas, they did not care. The name stuck. So soon, everyone was introducing me as "Texas" from Wyoming. When people my age or drunk folks asked me what I was doing in Australia, my mates instructed me to tell them "Just crushing tins and rooting!" which was quite an appalling statement to many people. Most people would gasp and then immediately be in tears of laughter.

The night went on and got rowdier and rowdier. It seemed by this time, almost everyone around the bar knew I was American and wanted to "have a yarn." One thing that seemed to be brought up often was that only Australians use the word "cunt" in daily conversation. In the USA, I would never even think to say the word out loud. It was one of my least favorite words, but somehow Australians just rolled it off their tongues.

Strangers began instructing me on the versatility of the word. "This is the only place where you call your mates a cunt and strangers' mate." Bystanders insisted I try using the word more. With whisky working as a wonderful catalyst, I greeted everyone I met the rest of the night with "How ya going cunt?"

I stood in a small group with my traveling partners, Chuck and Rhys. A young man who was quite drunk came up and, without using his hands, took a slurp out of Rhys's cup of whisky. Rhys was well known for having a short fuse. As soon as this kid did this, Rhys crushed the plastic cup, threw it into the kid's face, and stormed off through the crowd.

At first, we all laughed, but the young man who had done it had a sad look on his face. The look on his face turned to tears welling in his eyes. He finally said, "That guy was my hero growing up. I was just trying to be funny." A few people told him to chill out and just buy Rhys a fresh drink and let him calm down, which he did. It hit home to me at that moment how well-known, well-liked, and respected the guys were.

One man from our bull-catching crew had taken a few days off, driven down to the rodeo, and met us there. He was a massive man. We called him "Big Roo." He had played rugby most of his life but wanted to try some bull catching. Because of his strength, he was good at it. He had tattoos covering both of his arms. He was a nice guy but quite intimidating to folks who did not know him. As the bar closed, the police came in and started telling people they needed to leave.

The police instructions were met with resistance. Somehow, I got into a wrestling match with the biggest man there, Big Roo. I quickly lost the match, but he had picked me up and would not set me down now. He used me as a weapon, swinging me around wildly, using my legs to hit people. A big, fat cop was making people leave.

The cop came over to Roo and told him to set me down. The cop said we needed to leave. Of course, Roo would not do that. He used my legs to hit the cop and started making pig noises. I got a little worried. No matter how hard I struggled, he was too strong, and I could not get away. I thought for sure I was going to jail for this.

I had a belly full of whisky, so it slightly blurred my memory of whether this was a weekend cop or a real one. I remember him grabbing for either pepper spray or a taser and his handcuffs telling us to stop. Giggling like a little schoolgirl, the massive Roo ran, carrying me like a child. He easily outran the fat cop who was after us. I was worried but also quite impressed.

We then decided it was probably not wise to return after that ordeal, so we returned to our camp, a few minutes' walk away. Then, someone had the bright idea of visiting one of the other rodeo contestants' trailers for a drink. We did so, and they gave us full cups of homemade wine when we arrived. That topped the night off right, and we found our way back to our bedrolls as the sun came up.

24
Coppers

The shenanigans continued the next morning. Someone slapped me in the face with a cold can of beer, then kicked dust into my face. Of course, I sat straight up and began drinking the beer to wash the dirt out of my mouth. One more day of rodeo, and then we would drive to the next one somewhere I had never heard of, a few hours away. The day went as I had expected and exactly the way the days before had gone. We sat around drinking all day and meeting new friends. We sat around telling wild old rodeo stories, which was my favorite part.

After hearing enough of these stories, we went downtown to a bar. We stayed until almost closing time before getting a "Maxi Taxi," a large van to take us home. When we climbed inside, Rhys took the front seat. Almost immediately, someone asked for the radio to be turned on. Rhys did just that, and the taxi driver got upset. Rhys said, "Fine, I'll just sing to my mates," and pounded on the dash singing the New Zealand classic "Slice of Heaven."

The driver lost it as Rhys drummed on the dashboard. The driver turned around not even a block from where we started and dropped us off right next to some cops. The driver kicked us out of the taxi, and as we piled out of the taxi the cops made a few comments about "fucking cowboys" coming through their town. As they mocked us, they got closer, and a few people exiting the taxi had some things to say back.

I clearly remember someone calling the cops "shit-eating dogs" and telling them to leave us alone. They did not listen to this warning, and quickly a fight broke out. At first, the numbers were even. Then, it was a royal rumble. I quickly realized I had no desire to go to jail in another country, so I backed out. Most of the other guys did as well. But not Rhys and Chuck. These cops would not have them, so they continued to fight.

As I sat there rolling a cigarette, the two fought eight police officers without a problem. Finally, about six cops grabbed Chuck and got him to the ground. He giggled all the way down, reminding them all they had far outnumbered him and still could barely control him. As they finally got him in cuffs, they went to throw him in the back of the police truck. They had not loaded him before he kicked out one of their taillights.

Once inside, he continued to bang on the sides of the police car, exciting Rhys. They were like brothers. I knew Rhys could not let this go. Rhys asked one cop, "Would you let your best mate go to jail alone? Or would you go with him" the cop repeatedly told Rhys not to do it but continued to egg him on as well, pushing him in the chest. Rhys is not a big man, but you would have thought he was then. He rumbled with the police, and none of them could restrain him.

It took all eight officers to restrain Rhys. He was too strong and quick for them to put the cuffs on him or get him on the ground; this little man drug eight police officers around in the middle of the street. Finally, the biggest officer, twice the size of Rhys, got him in a sleeper hold, putting him down, unresponsive. People in the streets began throwing things at the police. They called the police names, letting them know they had won in a far less than a fair fight.

Rhys woke up, looked directly at me, and said, "What are we doing Texas?" I informed him, "Rhys, YOU are fighting the cops." Even in cuffs, he went on a rampage, requiring all officers to restrain him again before hauling them away. As they climbed into the car, I heard the cops say, "God, I hate cowboys, but I'll be damned if I fight one again."

The entire incident had started when the young officers, about our age, had made some comments. We could not turn and walk away from that. So, I walked to the police station to wait for them to release my friends so we could return to our bedrolls. I rang the intercom repeatedly, telling them to let Chuck and Rhys out until they threatened to arrest me as well. I then slept on the stairs of the police station until they let them out.

They gave them a notice stating they were permanently banned from this town when they were released. They could stay outside town at the rodeo grounds but never return. The police said the ban also went for "their friend who kept ringing the intercom all night, begging for his friends to be released," so I assumed I was kicked out as well. They commented that I was like an obedient dog, sleeping on the police station stairs; I would never leave them. I did not know if it was a compliment or an insult, but I did not care. We were all out, back together, and we would be off to another rodeo the following day.

25
Hangover

The following day was a rough one. First, we needed to drive a few hours south. This meant one unlucky soul would have to sober up finally after a three-day bender. Then, they would have to drive to the next town. I drew that short straw and had to drive the rest of the crew to the next rodeo. Chuck and Rhys sat in the back seat, laughing and telling stories and drinking beer. They continued the party without me. My hangover was almost unbearable. To make the hangover worse, this was the first time I had driven on a paved road in Australia. Since I learned to drive in the United States, all the lanes and laws were backward to me.

I had learned to drive a right-hand drive vehicle not long after arriving in Australia; they gave me the job of driving our massive bull truck. I had spent a lot of time speeding through the bush and flying down rough dirt roads while driving the lumbering, twelve-speed cab-over truck. The bull truck was much bigger, heavier, and more difficult to drive. But, there was a major difference: I had not been driving on a busy road with other people. Driving the bull catchers hadn't helped either. Bull catchers had to be driven incredibly fast, avoiding the anthills and trees while trying to to wrangle up a bull at full speed. However, that too had not involved other people or vehicles around.

Now I was bullied into driving. I felt quite sick as I drove that day. Luckily, because I was so nervous about driving, I had little time to think about feeling sick. Finally, the guys chimed in, asking how I felt. I responded, "I feel like shit." They told me, "Texas, I have just the thing for you. Have you ever tried a Berocca? This will make you feel much better. Just put it in this bottle of water and shake it up and drink it all as fast as you can." I felt like anything would help, so I grabbed the bottle.

I dropped in the large brown pill and shook. The capsule would not dissolve fully. It only dissolved partially. They told me to just drink what I had; it would be good enough. I took a swig of the bottle, and it tasted awful. I nearly spit it out. They giggled a little and told me I must slam the whole thing quickly because they do not make it taste good. So, I did just that, but I could not even swallow the entire thing because it tasted so bad.

Finally, I told them I had enough. I could not handle drinking any more of it. They began howling with laughter, and I could not figure out why. Maybe Beroccas were like Vegemite, something only Australians could love. But, on the other hand, maybe it was just the flavor. Maybe I would like a different flavor. It was not until they handed me a package that I understood why they were laughing.

When I looked down at the package, it had a picture of a dog. As I drove, feeling even sicker, I tried to read the package. What they had fed me was a dog deworming tablet. My head raced back to all the veterinary classes I had taken at the university.

I tried to remember if I could remember anything about dog dewormer being unsafe for humans.

I got more and more sick as I drove. They informed me it was a sure sign that the dewormer was working. They continued to laugh. They told me that surely I needed a good dose after living in the bush as long as I had, and maybe they were right. To make matters worse, they got rowdy when we entered the town where the next rodeo was. At every intersection we came to, they would scream. Instructing incorrectly. Telling me the wrong traffic rules. My mind was racing with anxiety as I tried to understand how to drive on the other side of the road. Each time I took a turn or stopped somewhere, they would begin yelling that another car was going to hit us.

When we finally arrived, I could not have been more relieved. We were in an even larger town called Mount Isa. This rodeo was one of Australia's oldest and most famous rodeos. There were people everywhere in town wearing cowboy hats and boots, gearing up for the big party. We found a small place on the side of the road, threw out our swags, and made a camp. Since this was a big and famous rodeo, cowboys came from all over Australia to compete. We went to a bar, and when we arrived where they were betting on horse racing.

I met some of the best cowboys in Australia. Some of them surprised me: they knew some of my friends from when they had ridden professionally in the United States. I even met one guy who remembered me from a college rodeo when I was riding for the University of Wyoming. Within a few hours, Chuck, Rhys, and I had found a new rowdy crew that would lead us into even more trouble. The fun had just begun.

26
Boxing Tent

Mount Isa proved to be nearly as wild as the first rodeo. Australians, at least the ones I was around most, seemed to love to fistfight. They liked it more than anywhere else I have ever been in the world. It seemed like everywhere we went, somebody wanted to fight a cowboy just because, whether it be the cops or someone in a bar.

Usually, at least one cowboy in the group they insulted was happy to slug it out with them. Rodeo cowboys worldwide seem to have a reputation for being both wild and double tough. This is probably even more true in Australia than in the United States. This likely drew the attention, making everyone want to fight. Everyone wants to be the toughest kid on the block.

Australians' love for a good scrap was never more apparent than when some of my new friends took me took the boxing tent. The boxing tent was a traveling roadshow that followed the rodeos and carnivals. They traveled to some of the roughest parts of Australia.

The owner and promoter of the show would stand high on a platform, banging a large drum to draw attention just before showtime. His voice rang out loud and a clear across the crowd as it gathered. He instantly had my attention, and the crowd outside the large red circus tent got huge. Next, the promoter began introducing his company and then introducing his fighters.

"This, my friends, is the last boxing tent on earth! My father owned it before me and my grandfather before him. I grew up fighting in this tent! I have had world champions step through this tent, some of the toughest ringers, miners, and rugby players you have ever seen! My men will fight anyone! ANYONE! And they will win. If you can beat them, I will pay you a cash prize. Now, who in the crowd thinks they can beat my fighters?"

He began introducing his fighters and telling stories about all of them. He introduced the fighters from lightest to heaviest. Some of them were golden glove boxers. Some were undefeated, and some were nothing more than tough guys down on their luck looking for money. Then, he began hand-selecting people from the crowd about the same size as his fighters in all the weight brackets. Men started raising their hands in the crowd, begging him to let them fight his hired fighters. He picked some ordinary-looking guys at first. Then the men he chose got meaner and tougher-looking.

He picked an Aboriginal man who claimed to have been a fighter in years past. Then he selected a skinny pale white kid covered in tattoos from Tasmania. Finally, some of my rodeo buddies had their hands up and got selected. One of them was a young man named Jake who had just turned 18. He was skinny and tough as nails. I had seen him ride in Cloncurry and win the junior events. He was exceptionally talented; he rode bareback, saddle bronc, and bulls.

The ringleader then picked the men in high-weight classes. He picked a massive Māori man covered in tattoos. He looked around for his other large fighter and could find no one his size. So, he offered to let two smaller men fight the big man simultaneous-ly. He selected another rodeo friend named Greenie, who I had bailed out of jail the night before for a separate incident, with Chuck's help. He was staying in our camp with us.

Finally, the show was on. We all paid our twenty dollars and rushed into the tent to claim the best seats. It was simply a small mat with some hay bales surrounding it. There was no ring or official gear, just two men with gloves on, toe to toe on a mat in the middle of a few hundred screaming drunk fans wanting to see blood and carnage. Somehow, I got separated from some of the people I knew. When I sat down, it was with five girls who had come with us from the rodeo. The girls were attractive, so they easily batted their eyelashes at some guys and got us front-row seats for the action.

In the very first fight, blood and beer started flying. Everyone cheered for the man just selected from the street. He was the underdog. Since there was no ring and we were in the first row, more than once the fighters would start falling toward us, exhausted or shaken up. All we could do was put our arms and legs up and shove them back to the middle of the mat. It was just a monetized, traveling playground fight.

The girl sitting with me was nicely dressed and had a white top. During one of the first fights, a man took a hard hit square in the mouth. Blood went everywhere. It splattered her white shirt. These men fighting were not professionals. Nobody knew if they had any diseases, so I covered the open top of my beer for the rest of the fights.

My young friend Jake was matched up with possibly the toughest fighter of all. He was not the biggest or strongest, but he was a golden gloves boxer, which meant he had fought in very high-end amateur boxing matches. He had grown up fighting in this ring. His Dad was an Australian boxing champion. He was back in this ring just to stay in shape for his professional matches.

I did not know Jake very well, but I heard from others he came from a rough area in Australia. He was all heart. To everyone's surprise, he stood in with this stud of a professional boxer. He was never once knocked down. He never backed up and even landed quite a few very sharp cracking punches on the professional. The fight went the distance, so a decision had to be made. The tent owner got the final decision, so his pro fighter won. I heard the pro walk up to Jake afterward and tell him he was the toughest street fighter he had ever fought. Of course he was: Jake was a cowboy.

Of the men selected from the crowd, only one street fighter won. It was an Aboriginal man who was half drunk, the one who claimed to have fought in his past. It was quite apparent when the fight started. He whooped the tent owner's man. All the other fights were at least close, and all the men from the street could fight well. I got the impression that any of the street fighters could have knocked out the other man at any minute.

Then came time for Greenie and a stranger to fight the big man. It was two-on-one, one massive man versus two smaller ones, but it was still not fair. The two men attacked the big man with all they were worth. They double-teamed him and took different angles, but they were no match. They stepped out of the ring with black eyes and lumps on their heads. Nevertheless, they had done their job and entertained everyone.

The boxing show was said to be the last on earth, and I can see why. It was dangerous and unorganized. It seemed like the next lawsuit could take down this business that had existed for generations. It was like a trip back in time. I was watching something you would see on the streets back in the 1800s. It made me sad to think it might be gone soon, but I was so happy I had time to see it before it went. Truly a night I will never forget of blood, beer, bruises, and a lot of respect being earned.

27
Mount Isa Short Go

The time at Mount Isa flew by. It was all a blur. Rhys made one of the best bull rides of the whole rodeo, getting him into the finals of one of Australia's most famous rodeos. He was the last man to ride. Everyone else had fallen off in the championship round. This meant that all Rhys had to do was ride his bull, and he would be the champion bull rider for the event. He knew this before he even got on. He normally just cruised through life. Even in stressful, dangerous situations, he was calm. This bull ride was different.

Rhys could already see himself holding the championship buckle and cashing the check for a few thousand dollars. He had an entirely different attitude now. It showed on his face. People quit talking to him because he had entirely zoned everyone out while deep in focus, his eyes honing in on one thing: the bull he was about to ride. He pulled his hat down as far as it would go to avoid losing it during the ride. Then, he popped his mouthpiece in and climbed into the chute.

I realized his laser-like focus had enveloped those standing around him who were preparing to assist him in getting on the bull, including me. Chuck pulled his rope, and I put a hand on his chest to protect him. It is standard procedure in bull riding to have someone spot you until you leave the bucking chute. The bulls often buck while still in the chute, which is extremely dangerous. I realized that there was suddenly a massive crowd of people around us. They hung from the stairs and off the chute and anywhere they could fit so they could watch the legend do his work.

Rhys did not waiver as the announcer notified the crowd that this ride was for everything. He continued to move with deliberation and meaning. He nodded his head, and they came careening out into the area. Everyone held their breath. Jump after jump he made the perfect move, staying aboard the bull. Then, in an instant, he had gotten slightly out of time with the bull. On the next jump, the bull whipped him hard into the ground as the buzzer sounded. The bullfighters worked to save him.

He jumped to safety on the chutes, turned, and looked back at the official clock. The clock said he had only ridden 7.6 seconds, not the required eight. The crowd booed, but Rhys and all the cowboys on the chute had seen it. He had fallen off ever so close to the needed time. Another man won the competition, and Rhys still made some money but did not earn the buckle.

Rhys was more upset about not getting the buckle than not winning top prize money. That seemed odd to me. Rhys told me stories about trading prize buckles for stupid things and cash when he needed them. Buckles meant nothing to him. He later told me he did not want the buckle for himself. Instead, he wanted to give it to me as a present, so I would never forget him. It was a shocking gesture to me.

We had one last night to celebrate, which we did, in a big way. Another big fight started so we moved to another pub to avoid more conflict. The bouncer stopped me at the door. He did not allow me into the pub. My ball cap had turned more and more sideways throughout the night as it got bumped. I still do not understand why this was an issue. Most Australians said it was common sense that having your hat crooked to the side meant they would not let me in.

At the last bar we went to, Rhys insisted we do an Australian tequila shot to end my last night with him. A group of about five guys did it with me. First, we shook a line of salt out onto our thumbs. Then, taking a lime in our free hand, they told me to snort the salt, squeeze the lime in my eye and then take the shot, and I surely would not taste it.

I thought they were kidding until they snorted the line of salt, so I did too. Then, laughing but groaning in pain, we all tipped our heads back and squeezed the lime. Rhys took his lime, squeezed it into my other eye, and handed me my shot glass. We all downed the shot and then yelled with pain and laughter. I could not see or smell for a few days after.

The rodeo ended, and that meant it was time to leave. For Chuck and Rhys, they would head back north to keep catching bulls. I was going to bum a ride south to catch my flight back to the United States. I packed up my things and said my goodbyes to my newly-found friends and my two buddies who had shown me so much. True to their nature, each gave me a hug goodbye, using the opportunity to give me one last rabbit punch in the gut for good measure.

I jumped in a car with two bull riders that were headed south. We packed our things in a little white Ford station wagon. We headed to the next bull riding event in a town called Rockhampton. The two guys I now traveled with were good bull riders named Burto and Lachie. Both men began riding in the PBR a few years after I left Australia: PBR is the most elite level of bull riding competition in the world. Later on, I would watch them on TV ride all over the USA and Canada. Burto eventually qualified for the PBR world finals two years in a row.

On our way to Rockhampton, we stopped in a little beach town called Bowen. We met up with another bull rider named Beau to do some practicing. Beau owned a handful of bucking bulls and had created a small practice ring behind his house. He lived in a recycled shipping container house he built himself behind his Mother's house. We spent a few days with him riding bucking bulls and going to the beach.

Few people can say they rode bulls and snorkeled the great barrier reef on the same day. These guys could do it daily. There was nothing professional about the practices. One man rode, and the rest of the people had to be bullfighters. Bullfighters step in and distract the bull once the rider has fallen off. If they were not riding, they fought bulls in flip-flops or were just barefoot. After our practice, the boys had a big rodeo in Rockhampton, so they dropped me off at the airport. I flew down to Brisbane, where I would spend a few days waiting for my flight home.

28
Bumpy Head

When I landed in Brisbane, I still had a few days to kill before my flight back to the US. I made plans to meet up with the girl who had originally gotten me the job in Australia. Her family lived not far outside Brisbane in a small farming town. She sent one of her connections to fetch me and bring me to the farm. On the farm, they raised cattle, sheep, and crops. I met the whole family, and we had lamb for dinner.

They gave me a small helping of sheep's brain to try; the first time I had ever eaten brain. The family told me about their history in the area, and I told them stories of home. The entire family was very well-traveled and had seen much of the US. I told them about my experiences out in the bush with their son, Lachie. Those were the stories they enjoyed hearing most from me. They loved hearing it from an American perspective.

I had met their daughter while she was studying in the United States. We both went to the University of Wyoming and had many of the same classes together. One day in one of our hands-on cattle classes, I asked her how hard it would be to find work in Australia. She replied, "All of you Americans ask that, but none of you ever actually follow through." I assured her that if I found work, I would go, so she gave me her brother's email, which was how the entire process started.

The family informed me that the following night there would be a big polo game in the next town. We were all going. We had rented a hotel room in town close to the polo grounds that twelve people would smash into. All my clothes were dusty and dirty from the months in the bush and on the rodeo trail. I had to borrow some clothes that were too small, but they worked. When we arrived at the polo pitch, the game had already started, and the party raged. The men all had on clean white pants and nice collared shirts. The women wore pretty dresses and enormous hats, like at horse races. I was glad they had forced me to wear fancy borrowed clothes, so I did not stick out like a sore thumb. They said everyone looked fancy but warned me after the sun went down that everyone would be blind drunk and act like fools.

They were right. When the sun went down, they lit a huge bonfire. Everyone was drinking a deadly concoction called Pimm's cup. I know it was full of fruit and probably had way more booze than anyone could tell. A band started playing, and everyone was dancing and having a good time. We danced until they were ready to kick us out late at night.

The last shuttle bus was getting ready to leave. The only way into town at this point was to take the shuttle, so we were stuck if we did not get on this bus. Throughout the night, everyone I met was friendly. They felt the need to buy me drinks and make me drink them. It messed me up like a soup sandwich. We tried to stay together because we only had a few hotel keys. I did not know where the room was.

The bus started driving away and people started chasing it. People jumped in through the windows and laid across other people to get a ride home. I made it on with two girls from my original group. They knew where to go. We had lost everyone else. When we came to our stop, they told me to get ready because we would need to jump off quickly.

I stood up, ready to go, standing in line right behind the girls. The doors opened, and they stepped off. I went to do the same when a big arm shot out across the aisle, blocking me from exiting. A big guy twice my size shoved me back. I became frantic. The girls yelled at me to hurry or they would lose me. I tried to squeeze by the guy, and he would not let me. His friends stood up from their seats as well.

I had been traveling and living with a crew of some of Australia's toughest men. I guess I had developed a chip on my shoulder between running with a tough crew and the booze, especially after all the fights there had been in the past few weeks of rodeos. I looked him right in the eye with his group of buddies behind him and said, "I bet you think you're a tough motherfucker, don't you?"

Famous last words. Without a word, the big man grabbed me by my throat, lifted me off the ground, and threw me flat on my back between the bus seats. A bolt sticking out of one of the old seats caught my ear, tearing it badly. After that, the big man and his friends kicked the tar out of me.

Some of the other people pushed them off the bus, and the driver closed the doors and sped away. He dropped me at the next stop and said nothing. I was alone in a town I had never been in before, with no phone, bleeding badly and totally lost at three in the morning. I began just wandering in the direction I thought the hotel was, but I was totally lost.

I found my way back to the stop where the girls got off, and the group of guys who had whooped my ass were still standing there.

I had blood all over my head and was certainly the only person with an American accent in that town. When I walked by, they asked if I was the guy they had just whipped. I just told them no, and they believed me somehow.

I wandered alone until I found a bench. I intended to sleep there and hopefully find my friends in the morning. A guy just leaving a nearby bar walked by and started talking to me. He was shocked that I was American and even more shocked at how messed up I was. He offered to let me sleep on his couch not too far away.

I walked with him, and we swapped stories until I heard someone yell my name. I looked up, surprised, and realized we were walking past my hotel. Luckily, someone looked out the window and saw me. I would have walked right past it because of the shape I was in. They came rushing out to get me and led me inside. I piled into bed with people sprawled all over the floor.

I had to fly home two days later. When I boarded the plane, I was quite the sight to see. I had my ear taped shut where I probably should have gotten stitches. I had a black eye and lumps all over my head from where I had been hit. I wore the cleanest of the filthy clothes I still owned, covered in Australian dust. I sat in my seat and daydreamed. I remember asking myself, *"Gosh, how lucky am I to get to live a wild life like this? Imagine the stories I will tell someday."*

Argentina

1
Headed South

Returning to everyday life in the United States was not the easiest thing to do. I had taken a serious disliking to wearing shoes, so I was always barefoot running around. I was rolling my own cigarettes, which had lost popularity in the US. It was nearly impossible to find loose tobacco other than in specialty stores. I threw the word "Cunt" around like it had no weight at all after a lifetime of resenting the word. Now I had to be careful when I let that word fly.

I carried with me a reckless lack of care for society's opinion or my safety. Those feelings seemed to be stripped from me when living in the bush. It was quite noticeable. Almost every single one of my friends noticed both good and bad things about my newly acquired behavior and vocabulary.

On one of my first days back in the States, my roommates wanted to go golfing at the local golf course. We brought bags of wine with us and rented golf carts. None of us were good golfers, but a few, including me, lacked all knowledge of golf course etiquette. By the third hole, we had lost every one of our golf balls. We still had six holes to play.

Without hesitation, I jumped into one of the water hazards full of golf balls. I came back with armloads of golf balls for my friends and myself. We finished the course using them. I did not wear shoes the whole time, and I sported shorts much shorter than normal in America. It drew stares from nearly everyone.

Many of my new habits were bad. They needed to be broken, and I did just that. Over the next few months, while I finished the final few courses of my university career, I worked on it. I quit smoking and started wearing shoes. I started acting like a more "normal" member of society.

I realized there was still one thing Australia had left me with, and I could never shake it loose. It had opened my eyes to just how big the world was and how different it could be. I had retired from bull riding one year earlier because of an injury. It seemed to have left a void in my spirit. There was still a yearning for the type of wild uncertainty bull riding had given me. I realized traveling had filled that void.

Years later, I told a friend about this and they said, "It's like you have 'l'appel du vide.'" This French phrase describes a phenomenon we know little about. It translates literally into "the call of the void." It is that sinister feeling a person has when standing on the edge of somewhere high. There seems to be some unexplainable gravitational force pulling you toward the edge, urging you to just hurl yourself over the side. It's the impulse you feel to jump before you can finally pull the brakes on your run-away brain train.

My friend meant this metaphorically, of course. I had no suicidal thoughts of hurling myself off a cliff, but I had felt the tug to danger often in life. Before starting to travel I often felt like I was slowly slugging my life away shoulder to shoulder with my peers like cattle being herded toward our "higher-paying" jobs. We would work for the rest of our lives in a mind fog until one day we'd realize the better portion of our life was somewhere far gone in the past. We had done nothing, and now death was near. That terrified me. My mind scrambled to grab onto something tightly. My heart inched closer to throwing itself off the cliff of societal norms and falling far into the abyss, where I may find certain death, or maybe certain life.

I began searching for another adventure. I was a junkie needing my next fix. It all came rushing to me one day when I was speaking to an old friend. My friend had a master's degree and still complained about having difficulty finding employment. Graduation was right around the corner. I was about to enter the job market; this comment was frightening. She had decided she was not getting hired because she did not speak a second language; she thought she needed to learn Spanish. We talked about how the language had become important in the modern-day United States. Spanish would only get more important with time.

Neither of us had any Spanish classes in grade school or university. Neither of us knew more than a few basic words in the language. She had decided that she was now too old to learn, and thus, she was screwed for the rest of her life. I would not give up that easily. I knew at twenty-two I would not be learning in a classroom. I had to do something else. The only other option seemed to be immersion.

I contemplated the idea of working on a ranch in a country that spoke Spanish. I researched all the places I could think of. Brazil seemed neat, but they spoke Portuguese. Mexico was close, but everyone made it sound terribly dangerous. Not knowing Spanish, coupled with the danger, seemed like a bad place to start. I did not want to go there, so I finally narrowed my search to Argentina. I began looking all over the internet for jobs in Argentina. I started emailing total strangers and making phone calls. Finally, after weeks of searching, I found an opportunity in Argentina. It would allow me to get my foot in the door. I made plans with the ranch or "estancia," as they called them in Argentina. I bought my flight for just a few days after Christmas.

I started telling my friends about my idea of traveling to a new land and learning a new language. They all shook their heads in disbelief but were excited to see what happened. We sat in my living room talking about the possibilities. Then, my room-mate's wife asked me a question that shaped my life. She asked, "What do you actually want to do with your life? Where is this all going?" I thought for a moment. Then, I just blurted out the first thing that came to mind. I told her, "I won't be the most interesting man in the world, but I'm going to be the most interesting cowboy in the world."

2
Immersion

The small estancia I had found in Argentina allowed English-speakers to work in exchange for food and a bed. This estancia had many English-speaking guests visit from around the world. They were one of the elite guest ranches for the international horse lover. They provided us with a list of things that we would be responsible for. These tasks included horse care, entertaining guests, some cattle work, and possibly helping in the kitchen. I figured I could do most of those, but I hoped I could stick with the employees who mainly worked cattle and rode all day. I booked my flight from Milwaukee to Toronto and then to Buenos Aires, where I would take a bus inland to the ranch.

My flight was in early January when the weather was not so great in the northern US or Canada. They delayed my flight an entire day because of the snow in Toronto. The Air Canada staff were genuinely nice and provided hotels for everyone who needed one. Canadians are famous for saying "sorry," and I believe in this airport, I heard sorry more times than at any other point in my life. There would not be another nonstop flight to Buenos Aires for a few days. They gave me a flight to Lima, Peru, where I would catch a connecting flight to Buenos Aires.

Arriving in Peru was the first major culture shock of the trip; nobody around me spoke English. We landed just as the sun went down, and out the window I could see all the tiny colorful houses near the ocean. Upon arrival, all the gates were closed. They told me I must sleep at the airport. The following morning, when I went to board my flight, they said the flight was full, and I did not have the proper ticket. So, there I was, stuck in Peru. The young girl at the counter barely spoke enough English to help me. I begged and pleaded to get on the plane, and luckily, at the last minute, someone canceled. They gave me the empty seat; I was finally headed to Buenos Aires.

In Buenos Aires, I had an old friend that I had previously worked with on another ranch. She was working for a travel company in Argentina. She invited me to stay with her and her boyfriend, who was a fantastic musician. I had tried to learn a little Spanish before arriving, but it had not worked, so I had a hell of a time explaining to the taxi driver where I was going. He eventually got me there after talking to my friend on his phone. The language barrier was proving to be more of a problem than expected, and it was only my first day.

It was incredibly uncomfortable traveling in a country where most people did not speak English. Asking questions was embarrassing. Sometimes not asking questions was even more embarrassing. I would end up ordering some food I did not like or walking long distances in the wrong direction. I missed my bus because I was scared to ask where it was going. It was exhausting. This is the part I remember the most.

They say immersion is the fastest and best way to learn any language. For me, it was not terribly fun. I now understood why people could learn so much quicker: they have no choice. It is mentally exhausting because the mind cannot be on cruise control. I had to rattle my brain to read every street sign or understand someone talking to me. I saw and heard words over and over. I knew I should remember. I knew what they meant, but my brain always seemed to be at full capacity for the day.

When I arrived at my friend's house, we went out and had the famous Argentine pizza. It was truly the best pizza I have ever had in my life. Buenos Aires had many Italian immigrants, specifically around World War Two. These immigrants brought with them their terrific love of food and wine.

Italians still have a massive impact on Buenos Aires today, hence the amazing pizza. We went to a liquor store and purchased "Fernet." Fernet is a strong alcoholic drink from Italy that tastes like black licorice. They told me it was the most popular drink in the country, other than the famous Malbec wine. All the young people drank Fernet, but there was a special way to drink it.

We purchased a two-liter bottle of coke, and then they took the coke and poured it into another container. They cut the bottle off above the red label, making it like a giant cup. They poured a mountain of ice into the bottle and then poured a hefty share of the alcohol. After it sat for a minute and the ice settled, they poured coke in until it reached the brim. The coke bottle became a large community cup. We then passed it around the table. We sat on the high rooftop and watched the sunset over the decrepit tall buildings in the distance.

People in Argentina really like to share drinks with an entire group, something that you rarely see in other countries. Argentinians did this with their alcoholic drinks and their hot drinks. Mate, pronounced ma-tay, is Argentina's favorite drink. Mate is a drink like coffee or tea. I would see people drinking it all hours of the day, but mostly in the morning, much like coffee in other parts of the world. They would pour hot water out of a communal thermos into a small cup full of herbs with a metal straw. Mate was the name of the cup, and "yerba" was the ground-up plant that was put inside. The cup was small, so it did not hold much water. One person would take a sip from the hot cup from the straw and then pass it to the person next to them. The next person would refill with hot water and sip. It was odd at first, but a tradition I came to love.

3
Medicine Man

I spent a week in Buenos Aires sightseeing and enjoying the nightlife. Since my friend's boyfriend was a musician, we got to sit in and listen to some epic jam sessions. In one man's homemade studio, we sat and listened to a musician recording. There is something wonderful about music when it is performed with passion. A person does not need to even understand what is being said to be entranced, to be taken into the world that the artist is bringing to life through sound. The band comprised of three young, incredibly talented women and one drunk old man.

The old man was the singer. He slurred, grumbled, and yelled words into the microphone even the native speakers did not understand. Yet, somehow, it sounded fantastic and full of passion. All the women in the group had brought their young children, all little girls. They sat on the floor and colored and played with whatever they could make into a toy. The little girls eventually fell asleep, unphased by the loud live music being played. Thick marijuana and cigarette smoke swirled around the room. Tunes bumped late into the night, long after we had run out of beer, and the night faded out of memory.

We went to dinner one night with a group of very educated young people who were our age. They all spoke English very well. We sat eating at an African restaurant of all places. They began asking questions about where I was from. Then the conversation turned to why I was in Buenos Aires and where I was going next. I explained I was headed out into the Pampas, to a large city called Cordoba. From there, I would go into the nearby Sierra Madre mountains to work on an estancia.

When I commented, "I am going to be a Gaucho," they all looked at me, puzzled. Finally, one man said, "Maybe you don't understand what that word means in our culture. I don't think you want to be telling people you're going to be a Gaucho." I had learned that a Gaucho is a South American cowboy. The word really had an entirely different meaning than I thought.

Gaucho does mean a South American cowboy, but they did not say it with the same pride as Cowboys or Vaqueros in North America or Ringers in Australia. These millennials from the biggest city in Argentina looked down on those referred to as Gauchos. Gauchos are an entire class of people, not just talented men working livestock with horses. I read somewhere that the word Gaucho originally stems from the word orphan.

Orphan made sense because they have no home. They are always on the move. I would later see that most poor people living in rural Argentina were referred to as Gauchos. This was a bit of a shock to me. All over the world, people who work in agriculture earn much respect from those around them; they are hard workers and providers. In Argentina, that respect just did not seem to be there most of the time.

With all this information overloading my brain, I was eager to get on a bus and make my way to find my new employer. I hopped on a bus in Buenos Aires. I expected it to be dirty, loud, and crowded, but it was the nicest bus I had ever been on. Each person had a leather seat that fully reclined, and they had leg rests so I could lie flat. I had my own personal television.

They served free meals and free alcohol on the bus for the ten-hour ride. It was incredible. I have never flown first class before, but I just pretended that was what it was like, and I was a king. It was an overnight bus ride. I sloshed around my glass of Malbec wine and finished my dinner, then passed out. I woke up in the bustling city of Cordoba.

The bus station was gigantic, with hundreds of large buses like mine parked along the sides, allowing people on and off. The next step was to buy a ticket to the little town of Rio Ceballos, where my employer would meet me. I did my best to make hand signals and use the Spanish words I knew to buy the proper bus ticket. Finally, they handed me a ticket and pointed me in the right direction. The ticket said the dock number and time the bus was leaving.

I stood where I thought I should be long after the scheduled time. I thought maybe the bus was just late, but it turns out I had missed the bus. They gave me a new ticket, but somehow I missed that bus as well. I was panicking. I bought a coke from a man selling them off his cart and sat down on the steps of the bus station.

It sunk in. This is real. I am far from home. Nobody is going to help me. My phone doesn't work, and I don't speak the language. I probably should not take out my phone anyway because someone will steal it. I just missed two buses in a row. There are no more today. I am screwed. The situation was like arriving in Australia. It truly tested my mettle. It's physically exhausting to carry everything you have on your back. The constant stress made the mental exhaustion even more powerful than the physical.

I sat there stressing out, trying to figure out how to contact my new boss to tell him about my misfortune. A drunk man sat down near me on the stairs. It was still early in the morning. He sat on the stairs, openly drinking beer and smoking a cigarette indoors. He turned and looked at me. He held out a can of warm beer with a shaking hand, offering it to me. I politely thanked him but said, "No, gracias," and he returned to minding his business on the stairs near me.

My mind suddenly forgot all my problems as I watched this man. He appeared homeless or extremely poor, and he was drunk in the morning, offering me a beer. My problems suddenly became so small that they became nonexistent. While this man's world was probably worse than I could imagine, he still seemed happier than me. I had spent heaps of money on this trip. I had spent hours of research trying to get here. Now I was so close, yet I sat on the stairs pouting like a little kid because my life was so hard. I decided I was going to fix that.

I stood up and began walking around the station with my heavy baggage, taking it all in. I read the signs and tried to understand what some of them said. There was a sign for the internet. I began reading in Spanish as best as I could about how to access the Wi-Fi. I would contact my employer through the only means I knew, email. After a few tries, I got it to work and emailed my employer, telling him about my predicament. Luckily, he saw it and responded immediately.

He sent a driver to fetch me from the station. As easy as that, I had solved my problem with a little "can do" attitude and some spiritual revitalization. However, as I sat in a different part of the bus station waiting for my driver, a dirty man with dark skin and dreadlocks approached me. He began saying things I did not understand but seemed to be begging for money. When I told him in Spanish, "Sorry, I don't speak Spanish, only English," his eyes lit up.

The man began speaking perfect English to me. He told me he was selling handmade dream catchers in order to buy his next bus ticket. He just needed to sell his last one, and he could buy the ticket. He was selling them for about two American dollars each. I did not have change, so I gave him around five, and he gave me his last handmade dream catcher.

I had nowhere to be, and neither did he. He just wanted to keep practicing his English, so we talked for a while. He had lived all over the world. He spoke six or seven languages, but he was originally from the rainforests of Brazil. Everything he owned in this world was in the small bag he carried on his side. He had traveled most of the world by living on the streets. He created art from things he found and cheap supplies. Then, when he had enough for a ticket to a new place, he just moved there.

I had so much admiration for the man. I admired his dedication to his teachings and his sense of adventure. He gave up everything to truly live a full life wherever he pleased. Just then, my ride found me and told me to follow him to his car. The Brazilian medicine man insisted he give me a blessing before I left him.

He told me, "This blessing will keep you safe in all your travels, but be careful. You may never stop, like me." He hugged me, looked deep into my eyes, and then turned and walked away. I often think back to this moment, wondering if he laid some kind of curse on me then, protecting me in my travels yet never allowing me to settle down. I wheeled around and followed the driver to his car. I jumped inside, and we headed to the estancia.

4
The First Estancia

Once in the car with my driver, we sped through busy city streets until we reached the outskirts of the city of Cordoba. We passed through an area filled with mansions. High-dollar polo horses roamed lush green paddocks. My driver did not speak English, so our conversation was limited. As we passed these houses, he said, "Polo, muy rico." I would later put together that he meant this is where the rich polo players and club owners lived. For the sport of polo, this was the Mecca for polo players and horses alike. The absolute best in the world lived here.

The farther we traveled from the city, the rougher the terrain got. We meandered up bending roads, higher into the hills. The hills were made of rough, rocky soil where crops would not grow. The only thing that grew there was the thick, woody under-brush. It was so thick it was difficult to see the soil beneath it. I wondered how cows and men on horses navigated the terrain. If the roughness of the terrain did not stop them, surely the brush would.

I saw entire families riding down the road on horses. In the little villages we passed, it seemed just as many people rode horses as drove cars. As the distance between us and the city grew, it seemed the people's wealth fell. So many buildings were falling apart. I got my first glimpse of entire families living in one-room shacks. Many young girls carried around babies. To me, they seemed far too young to be mothers.

We pulled into the long dirt driveway that led to the estancia. I felt an immediate sense of relief. I had finally made it. We drove our little car through a few streams and up the steep rocky road toward the headquarters. I had to get out and open about five gates to allow my driver to get his car through. We passed a couple of run-down houses with kids hanging out the windows. Guinea fowl scattered around the yards, and a few skinny dogs barked in defense of their homes. Finally, I opened the last gate. We drove into a lush green paddock filled with about twenty mares. The mares had beautiful, athletic colts running along at their sides.

We pulled up to the main group of buildings, and a middle-aged man greeted me. First, he introduced himself as the owner and the man I had spoken to on the phone. Next, he introduced me to the other English-speaking volunteers I would work with. All three of them were British girls about my age. Then he introduced me to his wife, who was also British, and we took a tour of the property.

Immediately, I realized he was quite stern with the rules set in place, and there were many of them. This was not surprising to me. All the other guest ranches I have worked at had the same structure. There were many rules and strict enforcement to keep a perfect appearance for the guests that came and went every week. However, none of the rules seemed to be too crazy, nothing I could not abide by during my time on the property.

They showed me to my room. It was far from where the guests stayed, down a hill. Our shack sat in a paddock full of the yearling horses that had been born on the property the previous year. The living conditions were primitive, nothing I was not used to. It felt like home to me. My room contained a small bed, a lamp, one window, and a dresser to put my clothes in.

The walls were clearly handmade. They were just rocks piled up with some mortar in between to hold them in place. The girls shared rooms on the other side of the structure. I shared my side of the small house with a young man working in the kitchen. He spoke no English at all.

We went right to work with the horses after I had settled in. I met the Argentine workers who were the guides on the horse trips. Only one of them spoke English. We were the same age. He was a retired polo player who had spent time in England playing professional polo. He had learned English while in England. His name was Adolfo.

The first ride on the horses was an experience for me. We used Argentinian saddles, which I had never seen or used before. Some of them were true traditional saddles. Others were old English or military saddles. I was excited to work cattle with the men, play polo, learn to rope with their rawhide ropes, and practice my Spanish. Little did I know I would do almost none of that during my time at this ranch.

They had instructed me to bring a horse-riding helmet, something I had never used or owned in my life. I thought maybe this was for the intense games of polo we would be playing. Instead, they required me to ride in a helmet every time I swung up on a horse. This slightly shocked me. There was no option. Two minutes into the very first ride, I was frustrated.

The local workers did not wear them, even the guests had the option to not wear them if they so chose, but I was required to. This did not seem to matter to the British girls I worked with.

They had all grown up wearing these funny things on their heads while riding. Half chaps, crops, constant pressure on the bit: these were things I knew nothing about.

When I expressed my discontent about wearing a helmet, the British girls laughed at me. Wearing a helmet made me feel like they did not trust my ability to ride a horse. It made me look incompetent to the guests I was trying to instruct and undermined my appearance as their experienced guide. The girls informed me that a ranch-raised American girl had left just a few weeks before I arrived. I had filled the position she vacated. They told me she had the same view on helmets for the same reasons.

For the three months she worked there, she complained about helmet rules, and the owner never budged. I figured voicing my opinion to him was futile. I would just suffer in silence. Then, I realized how petty of a thing that was to stew over. Every time I stepped onto a horse, the little piece of resentment came back. I would spend more time than I should have concentrating on the piece of Styrofoam and plastic bouncing around on my head.

5
The Kitchen Cowboy

My favorite part of working at this place was interacting with the guests. I was not a huge fan of being expected to cater to their every need. However, I absolutely loved swapping stories with them and teaching them about the horses. They came from all over Europe and North America, mostly. The guests usually came as solo travelers or in small groups, with one exception.

There was a company that offered overlanding trips across South America. One of their stops on the trip was the estancia. These were my absolute favorite people to deal with. Nearly all of them were my age. They were traveling solo and just experiencing South America with a group of like-minded strangers. When they came, we would take them all riding.

Most of the guests who visited were older and came almost exclusively for the riding experience. They were often tired by the end of the day. We would conclude our day with a posh, very proper dinner. The dinner had more pieces of cutlery than

I knew how to use. They gave us proper English etiquette lessons when we arrived. I assumed this was so we did not insult the guests with our heathen-like eating habits.

When the overland group came, it was the opposite, and I absolutely loved it. We had a massive barbeque outdoors that they call an "Asado." Live music, plenty of Malbec, games, plastic plates, taste-testing, tent camping, storytelling, and bonfires. It was just the thing I enjoyed.

The group that showed up with the overland trucks were the least experienced riders that visited us. There was much more work for all the staff. That did not matter though, because they were the most fun. They seemed to be the group that got the biggest feeling of reward when they stepped off their horses, exhausted. For some of the young adults, it was their first or second time riding a horse. They went up trembling with fear. A few hours later, when we helped them off their horses, they smiled from ear to ear, and they begged for more.

With the more experienced guests, we would do faster rides, galloping through the high-altitude grasslands, feeling the wind in our faces. We would ride down steep rocky canyons to private swimming holes, some of the nicest I have ever swum in. The local worker would always guide us from the front of the line. He did not speak English, so our job as volunteers was to strike up a conversation with the guests. Sometimes we translated if we could.

The local men were required to do all the dirty work that most people would not want to do. They had to be up before dawn and find the horse herd in the darkness. They brought the horses to the barn. They had to turn them out at the end of the day. They worked all the cattle and led the trail rides in silence. Guests would get off and eat or explore while the locals stayed with the horses. As volunteers, we would follow them. Within a few days,

I wanted nothing more than to be doing the job that the local men were doing.

When I applied to work at this place, I thought I would occasionally have to help wash dishes or something like that. That was not the case. They expected me to set the table to their high standards. I had to cook the eggs for breakfast. The eggs would be returned to the kitchen if not cooked perfectly. As volunteers, we had to serve with five-star precision, always taking and leaving things over the correct shoulders of the guests. Everything had to be done in the correct order.

We also had to tend bar at every dinner. This meant making cocktails and pouring wine, beer, or whatever the guests wanted. I began to seriously wonder what the hell I was doing there. I had come to learn horsemanship and stockmanship, and to explore the countryside to see how livestock were raised there. I had not come to learn hospitality or proper English etiquette when setting or serving a table.

I could not be too upset about the whole deal. I understood they were trying to run a five-star holiday experience. I had been told I would occasionally help with these things. I guess the owner and I had different definitions of "occasionally." When I voiced my opinions to my new friend Adolfo, he shared my disdain for the work.

Adolfo did not have to do the work in the kitchen or dining room, but he was sick of leading guests on the same trails daily. He told me he was leaving in about one month. When he left, he invited me to go with him to northern Argentina, where he was from. His neighbor was a veterinarian who owned a large estancia, and Adolfo would ask him if I could work there and learn.

I ran the idea of leaving early through my head. I had told the owner that I would stay for three months, but if I went north, I would leave one-and-a-half months early. The news would probably not go well. The owner had a no-nonsense, hardnose approach to managing the business. I decided to take some time to think about it all and see if the vet would even consider having me. Until I made that decision, I decided I had to make the best of my time at this place. There was still much to see and things to learn if I kept a sympathetic attitude.

6
Polo

While I was making my decision if I should leave or not, I discovered the game of polo. We would spend an entire afternoon playing polo with the guests on our private polo pitch. The game was slow and boring at first, as most sports are while trying to learn the rules. However, once I got the hang of it, it was exhilarating. Polo is a white-knuckle sport; horses and riders collide at speed and push against each other. Players swing heavy wooden mallets while galloping at breakneck speeds.

While polo looks exciting, it is really awkward to learn. The player must have both riding skills and enough coordination to connect their mallet to the ball. A solid whack on the ball can send it flying fast enough to break human and even horse bones. The horses and riders need special equipment to protect them from the deadly ball.

Polo, in my mind, had been a sport only for rich folks, reserved for the prim and proper atop their high-dollar horses, which they neither trained nor looked after when finished riding. Argentina changed this paradigm in my mind in a big way.

Argentina is the king of polo: Argentinians are not the inventors of the sport, but they are the masters of it. They rank polo players on a scale that tops out at 10. If you are a 10, you are among the top 20-30 players worldwide. Every 10 in the world was from Argentina when I visited there. Argentina even has its own equivalent of the world cup. It is simply Argentina against the world, and Argentina almost always wins.

People make different claims why this is the case. However, one fact was undeniable about children playing polo in Argentina: they all played it, rich or poor. I witnessed many kids on horseback break off a tree branch or find a broom; they would find anything that could work as a mallet, and they would ride along, practicing hitting rocks. Children hung off the side of their horses to make the perfect shot. There was something to be said about these young Argentinian players and their tenacity.

For many, polo was not just a rich man's hobby: it was their only way out of poverty. They played with ferocity, as this was how they fed themselves. If they played poorly, they had no money to eat. This was the case with my friend Adolfo, and how he ended up playing polo in England and learning English. A wealthy man brought him to his polo club, and the better he played, the more money he made.

I never played in anything close to a professional game. I never got to witness the high-end polo matches played in Argentina. That didn't bother me, because the local small-town amateurs were just as fun to watch and play with. The same risk was involved in all games, whether they were high-end or amateur. Serious injury to the horse or rider was very possible.

I spent my first day on the polo pitch at a walk, trying to hit the ball off my patient horse. When I got the hang of that, I tried hitting at a trot. Bouncing along on a hot-headed thoroughbred trying to hit the ball at a higher speed took some practice. The more experienced players told me not to worry. They said it got easier to hit the ball as you went faster, and the horse's motion became smoother. I tried running and hitting the ball, and they were right. Faster was easier. However, it seemed like you could play for a lifetime and still have work to do to become better.

There was another far more brutal game they played that fascinated me. The game was called "Pato," which translates to "duck" in English. It originated as a game between competing estancias. Their hired men would place a freshly-killed duck on the border between ranches. Each team would start at headquarters and race to get the duck. Players could not get off their horse to pick it up, and whichever team got the duck home got to eat it. In modern times, the game is played using a soccer ball with handles sewn onto it, but the concept is the same.

Pato is similar to polo in many ways. In both games, players must lean down off the side of a galloping horse. In Pato, players hook their toes in the saddle to stay on top and reach to the ground to grab the ball. Once a player has the ball, anyone can grab it from them. If two players ride in opposite directions while holding on, whoever falls off first loses. It is a brutal game of tug of war aboard thundering horses running at top speed. People are almost always injured. I, unfortunately, never got to see this game played in person; I only learned about it from others. I did get to practice reaching down off my horse to attempt to grab the ball, but I could never seem to grab it.

Outside of soccer, most Argentinian sports seemed to revolve around the horse. Argentina is a country quickly trying to catch up with other powerhouse countries in our technology-filled world. This was obvious when I spent time in the bigger cities. However, when I got away from the hustle and bustle of the humming, polluted cities, I could see that people still lived and died by their horses. Locals were still very much attached to the equine creatures they both worked hard and played hard with.

7
Big Changes

I had been at the guest ranch for nearly a month now. It had offered me small glimpses into the rough life of a gaucho. I saw it through a looking glass that filtered out all those things a middle-aged woman on an equestrian holiday may not want to see. Finally, a gaucho holiday was coming up, and we planned to take a visit to the nearby church with the guests from the estancia. A ceremony would take place. People from the hills all around would ride their horses in their finest dress and horse tack.

Hundreds of gauchos showed up for the holiday at the Catholic church. Guests were in line behind us as we rode our horses to the festival. We tied the mounts to trees on the edge of the small encampment. The estancia owner loaded the guests and guides into a truck and drove us to the church so the guests would not have to walk. Riding in the truck was so awkward. We honked and drove up the street with white tourists hanging over the side, snapping pictures. Everyone stopped what they were doing and stared as we passed.

A large parade filled with people, young and old, all on horses, rode into the open area in front of the church. They wore colorful traditional dress. The men wore wide-brimmed hats and sported bright shirts and the famous riding pants of Argentina called bombachas. The women wore beautiful long dresses. Most of the dresses appeared to be handmade. Many of the girls rode sidesaddle to accommodate their dresses. Some rode in the normal fashion, for they were not about to miss this celebration. We watched as they tied up their horses and filed inside the old Catholic church to attend a short mass. I sat outside on a stump listening to the barely audible priest speaking through a PA system.

People were selling drinks, souvenirs, and ice pops at stands all over the place. The guests wandered around taking pictures. Some tried to interact with the locals, though none of the guests spoke Spanish. One guest joined me where I sat in the shade. She was a Canadian woman who had done well for herself by founding a healthy food company.

We spoke about the ceremony and shared thoughts about how we wished our Spanish was better. Maybe then we would better understand what was happening and why they were celebrating. She too had felt uneasy about being the only tourists there. We were not in danger, but we stuck out in the crowd. Even though we felt out of place, the trip had been worth it: the ceremony was a beautiful spectacle.

We returned to our horses not long after, mounted up, and rode home. The celebration would last a few more days, but we were told it became dangerous at night. The alcohol started flowing, and many gauchos were notorious for drinking and fighting. Later that night, I spoke with one of the local men I worked with. After the day's work on the estancia was done, he had ridden back to the event. He filled me in on what he had seen. Just as promised, there had been a lot of drinking and fighting, but overall, he said it was a lot of fun.

The problem with fights in these communities was that they usually involved more than fists. People sometimes died. Gauchos carried around huge knives about a foot long in their belts. They used the knives for work often. When they got drunk and started fighting, the knives could easily be pulled, and it would become more of a sword fight. Gauchos also carried a "rebenque." A rebenque was like a crop or short whip. They used it in working cattle and riding horses. If the knives did not get pulled into a fight, the rebenques sometimes did. They would whip the hell out of each other. The rebenques often had small metal rings or studs braided into the rawhide, which could do even more damage. The local employee who had returned said he saw a fight like this, and one man was badly hurt.

The next day, after returning to our little resort's safety and quietness, we went back to work as usual. However, after seeing the culture, the horses, the people, and the traditions at that ceremony, I became more perturbed with my situation. I could not stand doing kitchen work, serving dinners, and entertaining foreigners. I'd had enough of wearing a helmet and hiding from the real Argentina. For days, I'd thought about how much I wished I could just ride with the gauchos, to live with the men, not deal with the tourism.

I had bitten off more than I could chew. I had never had wonderful luck working in the equine tourism industry. It was just too structured and safe for my liking. I spoke with my friend Adolfo to make sure the offer to work for his neighbor, the veterinarian, still stood. Adolfo said the offer was still open and he would happily take me there, so I set up a meeting with the owner of the guest ranch.

I told the owner that my current job was not working out and I needed to do things more my style. I had not come here to be a hospitality worker. Originally I had agreed to stay three months, but now I was leaving after a month and a half, so he was upset. Adolfo would not leave to go home for another two weeks, so I told the owner I was happy to continue working until I could leave with Adolfo. It would give him time to replace me.

The owner said he needed time to think. His face was pulsing red with anger. He called me back to his office less than an hour later, telling me I needed to be off the property the next day. He said, "If your heart is not fully here, I want you gone," and so it was. I later learned his reasoning: he did not want me to spend two weeks telling the other employees or guests how much I disliked this job, which I could understand.

I felt I had done the right thing by telling him I was leaving two weeks in advance. Part of me wishes I had kept it a secret until the day I was leaving and then just left, but that would have weighed on my conscience for years. In retrospect, with time to filter out the emotional thoughts, I can look back and see why he did what he did. I am pleased I handled it the way I did. I can live with myself.

Once I left the guest ranch, I would be stuck in the city of Cordoba for two weeks until Adolfo could pick me up. I knew nobody there; I knew nothing about the city. I once again was alone in a country where I still did not know the language very well. I asked the taxi driver to take me to a cheap hostel, and he did. It turned out to be one of the best and worst experiences of my life.

8
Thief

When I was dropped off at the hostel, I did not know what to expect. I had little money to spare and two weeks to kill. This was the first time I had ever been to a hostel, and I had heard very mixed things about them. Many of the rumors were untrue. In the United States hostels are not common, so the people I had heard talk about them had never stayed in one themselves. Still, I was under the impression that they were grungy: full of drugs, disease, and down-and-out people. How wrong I was to expect this.

I walked in and was met with a big smile and someone who spoke English. They began explaining the prices, which were unbelievably cheap to me. I was expecting hotel room prices, $40-50 USD, and I got a room for less than $10 USD. I chose a four-bed mixed dorm with a private bathroom. This was considered one of the nicer room choices unless I wanted a completely private room.

The front desk told me about all the free activities they offered. They had a full kitchen, and the supermarket was just down the street, so I could sustain myself on cheap food. They had Wi-Fi, so I could call or text back home whenever I wanted. This situation that had me really bummed out suddenly had become a lot of fun. I went up to my room to set my things down.

Just as I was getting settled in, the door opened. In walked a beautiful young lady carrying an armload of baggage. I assumed she did not speak English like most people I had met so far, so I smiled and waved hello. She said, "Hello! How are you?" I was so relieved she spoke English. She had a thick accent that I had heard before but could not place.

She began telling me she had just arrived from Spain. Her name was Silia, and she was studying to become a doctor at the Catholic university in the city. Like me, she did not know anyone in the city, so she ended up staying in this hostel as a last resort. While she was staying here, she planned to search for an apartment to rent while she went to university. We spent quite some time making our beds and talking about the events that had led us both to this place. Then, we went to get food and explore the surrounding neighborhood.

We became quite close; we did everything together the first few days because neither of us had made any other friends. Silia was my lifeline. She translated for me whenever I needed it. Only the two of us occupied our four-bed dorm for three or four days. One day, we returned from dinner, and a man had moved into our dorm while we were gone. We thought little of it, grabbed the things we needed, and quietly left the room.

We met some new friends, other people studying at the same university as my Spanish companion. They were foreigners from all over Europe. We made good friends with a German guy that was my age. He spoke fluent English, German, and Spanish. We all went out and had some beers that night. Then, we even went to the club and danced until the light blurred and the music muffled. We stumbled back home.

When we got back, we quietly slipped into our beds. Our new roommate was sleeping in the same room, and he had passed out. Around four or five in the morning, I rolled over to see our new roommate walking out the door. When we woke late in the morning, we went to the courtyard. Silia had some things to do for university, so I went to find something to eat.

I needed to pay for a few more nights at the hostel, so I went into my bag to find some cash. The street exchange rate was about fifteen pesos for one dollar. The bank would give you only ten pesos for one dollar, so I had brought just under one thousand dollars cash with me. I had used about two hundred of it so far to travel in buses and taxis.

I was planning to grab one hundred dollars and go swap with a street dealer. I felt nothing when I stuck my hand in my bag where I had hidden my money. I was rattled and began going through everything I had. Still, I could not find my cash. It was all the money I had, and I panicked. I called Silia and told her what had happened. She returned to the room in a frenzy, with tears about to roll down her cheeks. She began accounting for all her expensive things. They were all there, luckily. We went to the front desk and reported to them what had happened.

They had cameras all over the place, and every person who checked in had to scan a passport or legal document. They knew exactly who everyone was. We started watching the camera footage. We discovered that Silia and the new guy were the only two who had gone in or out of the room during the time frame in which the money went missing. The new guy was from Argentina, a local, and they had his information in the system.

I asked them to notify the police. The man had left early in the morning and had not returned yet. He had left his things so he would probably come back. They looked at me with pity and informed me they could call the police. They told me that unless I bribed the cops about as much as had been stolen from me, they would do nothing. The cops were too crooked. Here was my introduction to staying in a country where calling the police does nothing.

We argued and pleaded with them, but there was nothing they could do. The money was gone, and it was probably gone forever since the man had left the hostel. The owner even spoke to me and apologized. He told me this was the first major theft they had in this place. He felt terrible.

The owner let me know he could not replace the money. I had signed a waiver when I arrived. They were not responsible for things like this. It was one of the lowest feelings I had ever had in my life. I think it was noticeable on my face. I had made friends with nearly all the staff during my stay and with many other people staying there. People went out of their way to give me hugs and tell me what a scumbag and piece of shit the guy was for doing that. A few days passed, and I gathered up change and a few bills the thief didn't find. He had even taken most of my bank cards, so I could not even have someone put money on my account in the US. Thankfully he had missed one card, so I made plans to get it working.

9
The Thief Returns

I spent the next few days worried. I tried not to spend any more than I had to. I spent most of the time figuring out how to get some money, if possible. I was lucky to have Wi-Fi, which made it possible to call home and help get things sorted out. When he fled, the man who stole my money had left most of his stuff at the hostel. He did not pay for his room, so they took it all and put it somewhere safe. He came flying into the hostel days later, high out of his mind and looking for a fight.

He had spent all my money on drugs. It was all gone. Now he wanted to fight with the staff and, surprisingly, Silia and me. I was furious. I wanted to kill this guy, but I knew that would get me kicked out of the hostel. Plus, he had done crazy amounts of coke and maybe other things that day. Fighting someone who can feel nothing is dangerous. The other problem was that I did not know who he was connected with. He could have easily paid some thugs to jump me if I left the hostel.

He marched over to us, a group of at least ten sitting around a table talking. We were talking about what a piece of shit he was and smoking cigarettes when he took us on. First to jump in and stop him was one of the hostel workers, a man I had made exceptionally good friends with. He was a huge Haitian man. Everyone else was ready to happily jump in and whoop him if given a chance. I think he realized that and started a verbal confrontation instead. He spoke no English, and I spoke little Spanish, so whatever was being said, I could not understand it.

Fiery Silia jumped in to give him an ass chewing. I did not even understand what she was saying, but I could sense it. This was the first time I could really feel the heat of language. I felt the passion in the body language and pronunciations. When you understand all the words, you discount some more subtle parts of speech. However, when you do not understand any words, you can understand only tempo, anger, and body language.

This man denied taking my money. He said something about me being a stupid foreigner anyway. He then turned and accused Silia and me of stealing a phone charger and a flashlight. He said we had taken some other insignificant items that were meaningless compared to the cash he had taken. The argument went on between the scorching Spanish girl and the coked-out man. The Haitian man held him at bay.

Standing behind the Haitian was our German friend, and I stood next to him. If the cokehead got through us, eight Argentinians stood chomping at the bit, ready to put him down. Then, finally, the man left. Even though I had not understood most of what had happened, it left me quivering with hatred, sadness, and confusion. I was so let down that assholes like this guy existed in this world. I was furious that nobody could do anything about it. I was feeling bad for myself that I had let this happen.

For days afterward, I caught myself saying, "You can do this, one day at a time." For the first time, I realized I was talking myself through fear. I was thinking through fear. I had spent a large part of my life surrounding myself with fear. I craved the adrenaline that came with it. I rode wild horses, raced dirt bikes, crashed cars at high speed, and rode bulls during my childhood. I always threw caution to the wind for a thrill.

Until now, fear was something I had reacted to quickly and thoughtlessly. This was the first time I was thinking through it. My parents used to read me a book as a child about a little train trying to climb a mountain. The train grunted and groaned, chanting to himself, "I THINK I CAN, I THINK I CAN, I THINK I CAN." For some reason, in my adult life, I often found myself saying this in my head. This was one of those moments.

John Wayne said, "Courage is being scared to death, but saddling up anyway." Nelson Mandela said, "I learned that courage was not the absence of fear, but the triumph over it. The brave man is not he who does not feel afraid, but he who conquers that fear." These quotes rang loud and clear for me. I could not avoid being afraid. I had to learn how to be good at being afraid.

With that in mind, I decided I could not dwell on what had happened. Instead, I figured out how to get a little money on my card. I eagerly awaited Adolfo so I could go back to work. Until then, I had to have as much fun as I could. I practiced my Spanish daily. Many of my new friends helped teach me. I used a language education app on my phone. The app taught me more in just a few days than I had learned in conversation in almost two months in Argentina.

I could go out on my own and order food or buy things at the grocery store and understand simple conversations. This was immersion at work, the reason I had come to South America in the first place, and it was paying off. The harder I made it for myself by getting into situations I could not back out of by speaking English, the faster I learned Spanish words and phrases.

10
Street Dogs and Babies

Adolfo would arrive in a few days. I conserved my money by only eating cheap hotdogs and bread people had left at the hostel when they moved on. I used a little mayonnaise to help the dry, bland concoction of meat and bread down my throat. University had started now. All my new friends spent their days on the campus. This left me by myself to wander the city or read or do anything I could find that was free. I learned a lot about the people of Cordoba during this time. I can confidently say that Argentina's girls are the most beautiful girls I have seen anywhere in the world.

Sometimes I would just sit on benches and watch people. Families would bring their children to small city parks and squares and sit all day. This is probably normal in big cities, but it was a first for me. It was mind-boggling that the only things that attracted these people to sit here, content all day, were green grass, trees, and a little elbow room. It was shocking to me because, in my line of work, I had more of these things than I could shake a stick at all day, every day. I decided maybe I took for granted what I had. I watched these dwellers of the concrete jungle congregate on the only grass for miles around.

Another thing I found fascinating was the street dogs, how they had a hierarchy and how they survived. I have a bachelor's in animal science, so my mind wandered to Darwinian theories about what made a great street dog. What characteristics allowed certain dogs to thrive and others not? I never saw a dog that looked purebred. They were all mutts. This could have been because the crossing of breeds made better dogs. It also could be because purebred dogs got picked up, and people kept them. I noticed that there were no incredibly large dogs or tiny dogs. None of the dogs were aggressive toward people. I think large dogs required too much food to survive. Small dogs could not protect themselves. Dogs that were aggressive toward people were probably killed or beaten until they ran far away.

The dogs were incredibly smart. Some of them would only cross at the crosswalks when people were crossing. I wondered how many of their dog friends they had watched perish by car before they figured that out. So many had three legs, one eye, or gnarly battle wounds. Yet, despite living a life much harder than dogs where I came from, their tails still wagged the same.

I saw this in the people as well. Then, one day, I saw something that will forever be burned in my head. When passing down the quiet side streets, I often saw poor people sleeping in boxes on the street. These people slept on the buildings' ventilation grates to keep warm at night. They jingled a cup full of change during the day.

When walking down one of these streets, I saw a child, still too young to walk, playing on the edge of the road. The child sat in only a diaper playing with an empty cigarette box. The child crawled around the dirty city gutters, finding unique pieces of trash and playing with them. A man with one leg sat with his back to the concrete wall nearby.

He was not dead, but he was clearly so drunk or high he could not move. Next to him sat a woman, I presume the child's mother. She too had glazed eyes, but she kept smacking the man and cussing at him for something he had done. I looked back at the dirty, naked child who just smiled, playing with the trash toys. He smiled as if the world was a wonderful place.

It was a situation that weighed heavily on my mind. I wondered if there was something I could have done. I wondered in how many other places around this city, around this world, this was happening. People like me walked by, unsure of what to do. The city continued to hum around them. The buses spewed dirty black smoke, and the surges of people going about their busy lives continued to flow down the major streets. People moved along like ants gathering food to bring back to the colony. They did not see or care what happened on these side streets. People preferred it that way.

Walking these roads less traveled allowed me to see what was really happening. It filled my mind with images I cannot remove. It was as if each day, my stroll to nowhere was a short walkabout or vision quest, a version of the rite of passage that transitions a person into adulthood. Suddenly, my stolen cash became a minor problem for me. I was upset with myself that I had developed such an attachment to things when this was how many people in the world lived, a world of dirty babies playing in the city streets with cigarette boxes, smiling with a few teeth.

I returned to the hostel that day with a different worldview. Someone complained that their phone cord wasn't long enough to reach the bed. I could not get that baby out of my head. I sat with a group of guys from Buenos Aires on vacation, and we talked. We watched the tall employee from Haiti serenade a beautiful French doctor. He was teaching her how to tango to electronic jazz music. We passed around a joint filled with awful, half-wet stem-filled marijuana that made my tongue numb. The nights were hot and humid. Finally, I floated back to my bunk bed and drifted to sleep.

11
Waterfall

A large crew of my friends from the hostel and I took a trip to a very odd town in the mountains nearby. It was a village that, despite being in Argentina, was all German. The buildings, the signs, and all the businesses are German. It was, of course, a tourist trap, but it was not overly busy, so it was not bad. We stayed in a cheap hostel that overlooked the valley below. You could rent horses at the bottom of the mountain to carry your things up to the village. They did not lead the horse or tell you where to go. They simply handed you the rope and said to have him back in a few hours. I had developed the identity of the American cowboy among the group. They all joked and thought about renting a horse to carry everyone's bags, but we decided against it.

The main reason we had been drawn to this funny little town was not the odd knick knacks for sale or the German food. There were rumors of an incredible waterfall nearby with a swimming hole below. We went out and had a nice German-themed dinner that night, then got a good night's rest so we would be ready to hike to the waterfall the next morning. We had been told it was not terribly far, but the trail was rough and hard to get down. It would require balance and some rock climbing.

We left early, stopping at the local store to buy a few bottles of wine. We packed some food for a light lunch, and we set off up the trail. I had grown up in the mountains, and I had hiked trails much steeper and longer than this with my father when I was young. I could see that some people had fallen to the back of the line. Apparently, they had spent little time outside their flat cities.

The trail twisted through low, wet undergrowth. It required a person to bend far forward and walk like that for minutes at a time. The ground beneath was slippery and muddy. Large rocks scattered the trail. We came across a stray dog. He was interested in us and made friends with some girls at the back of the line. He kept following us as we hiked along. Soon we could hear the roar of a waterfall. We knew we were close.

We were unaware that the trail's roughest part was still to come. When we arrived at the bowl, we were still high above the waterfall. We would have to work our way down into the bowl, which meant steep edges and downclimbing. Our group had quite a few folks from Spain, a few Germans, and a couple of sorority girls from somewhere in the southern United States.

When we came to the steep rocky downclimb, people's ineptitude for the mountains really showed. I heard a few girls yell from above. They were scared to break their wine bottles. Another complained of possible injury if they fell. These were both legitimate concerns. We came up with a plan. The people who felt more comfortable on the rocks would help carry down the packs in a few big loads. The others would stay back at the hardest parts of the downclimb to assist those who had a hard time. Like a herd of turtles through peanut butter, we made it down to the bottom.

The waterfall was one of the most beautiful I had ever seen, well over a hundred meters high with a cold, deep blue pool below it. A few local families were swimming at the pool, and that was all. We had all worked up a sweat getting to this point and could not wait to hit the water. Some of us stopped to put on sunscreen and change clothing.

I dove right in. I was tan enough from working outside that I did not need protection at this point. I had given up on trying to take care of my tattoos by protecting them from the sun. When I hit the water, my testicles jumped into my throat. It was ice cold, almost too cold to enjoy. I was one of the first in the water. I did not dare to comment on how cold it was because I feared the girls would not jump in.

We spent the day swimming, drinking cheap Malbec wine, and exploring the nearby area. Everyone got a little drunk and sunburned, but it was truly one of my favorite days in Argentina. I was surrounded by great friends and beautiful girls, far in the middle of the South American mountains. The scorching sun beamed, and cold water flowed out of the mesmerizing waterfall. I did not have a care in the world. For a day I forgot about the rough time I had been having the past few weeks.

Hiking back out of the waterfall was slower and more dangerous. We had to get out of there before the sun went down, and we were lost. Everyone was a little tipsy now. Most people had not brought proper hiking or climbing shoes, and now we were going uphill. Again, at a glacial pace, we slowly got everyone up the rocks without issue. We went back through the tunnel trail grown over by brush to a wide, flat trail where nobody would have issues. We barely got back in time. The sun was just setting, and we were all exhausted.

The following morning, we woke, walked back to the bus station, and rode a few hours back to Cordoba. Silia and a German friend named Adrian had found an apartment. They moved in there with another Spanish girl. They invited me to stay with them so I would no longer have to pay for a hostel room. The flat was tiny: one bedroom with four people in it.

The girls took the bedroom, and Adrian slept on the fold-up bed that took up almost the entire tiny living room. The tiny couch was where I slept. I am not a tall person, but it was far too short for me. I took the couch cushions off and laid them on the little bit of vacant floor that was left. It was not comfortable, but it was free, and I was with my friends.

We had fantastic conversations about European politics and life. It is amazing when you travel how in a short time frame, you can make so many meaningful friendships with people you have only known for a week or two. Friendships sometimes more meaningful than friendships I have had for years. Adolfo showed up at last. I said goodbye to my fantastic European friends who had been by my side for so much, to whom I was so thankful. Then, I boarded a northbound bus with Adolfo.

12
Reggaeton

Late at night, Adolfo and I arrived in a tiny town close to the northern border of Argentina, an area called Corrientes. From some of the tall hills, you could see into Brazil. We took a taxi to Adolfo's mother's house. Since it was late, we snuck in and went straight to bed. The following morning, I woke up to the smell of breakfast cooking and steaming mate.

Most people I met had a special cup to drink from. They kept the cups for decades. The old men often drank out of the cups they used as kids. Some of them even had the cups their fathers had used. Everyone carried around an insulated bottle of hot water and drank it one small cupful at a time. Adolfo's mother repeatedly told me that my mother was too far away, so she was my mother in this country.

Adolfo showed me around the town that day. His family owned an estancia far outside of town, but he said I would learn far more with the veterinarian. We had some time to kill because the vet I would work for wouldn't come until the next day. The town was like many towns in Latin and South American countries. There was a large square in the center, where the Catholic church and other important buildings were located.

Northern Argentina was much hotter than the parts of Argentina I had come from. I was now right on the border with Brazil. I could see that based on the foliage; they received much more rain and heat up here. Big, green broadleaf plants grew everywhere. The timber industry was particularly important in this part of the country. I saw logging equipment everywhere. The trucks hauled huge loads of trees through town. Their heavy loads were barely held on by one chain. The falling logs would likely kill anyone too close to the truck if the chain broke.

We met many of Adolfo's friends that he had grown up with and had pizza for dinner. He said we would go out with his best friend later that night. His best friend's parents owned a vacation home on the border of Paraguay in a town called Posadas. Adolfo told me his friend had a lot of money, but I was shocked when we got to his house. His house was something out of a hip American or European city: top-of-the-line modern everything, which was a bit shocking next to some of the shacks his neighbors lived in. In the driveway sat a brand-new Mercedes.

We loaded into the car with a few beers and headed to the city. As we drove away, he pulled out a button, and when he pressed it, his entire house flashed like a car and locked itself down. They called up all their friends to see what was happening that night. We flew through the night in the black Mercedes. The car was well built and high tech. I did not notice how fast we were going until I looked up front and saw we were going well over a hundred miles per hour.

The closer we got to the city, the faster he drove, and the tires squealed around each corner. I felt myself tensing up, but I realized there was nothing I could do, so I closed my eyes and took big, deep breaths. There I was again, calming myself down, thinking my way through the fear. I noticed how beautiful a monster fear was and how I could learn much about myself when I controlled it.

It was getting late, almost midnight, but for people in Argentina, the night was just starting. When we arrived, we went to dinner with some friends until about two in the morning, and then we went to a bar for a few beers. The nightclubs had just started opening at three in the morning, so we walked to a nearby club. Reggaeton music filled the air. We could see the closely-knit crowd of people moving, rippling like a rain-filled puddle.

It seems that people have really developed a disdain for the smell of sweat when it comes to sport or work. But, when the sweat is pouring out of people on the dance floor pulsing to the sound of reggaeton, it's different. Sweat is nothing but lubrication for the ebb and flow of a giant organism made up of people falling in love with strangers.

A girl from across the border in Brazil had taken a liking to me, grabbed my hand, and pulled me onto the dance floor. I could not understand her Spanish because of her accent, and she spoke no English other than the words, "LET'S GO!" so we did just that. She dragged me around the nightclub for the next few hours, yelling, "LET'S GO" and introducing me to her friends. We could not communicate.

Between the loud music and our awful accents, there was nothing to be said, only dancing to be done. Finally, the club turned off the music and told us to leave. We walked outside, and the sun was coming up. Everyone was hungry, so we went to a nearby restaurant serving breakfast. It was full of people getting up and ready to work. Everyone else was like us. They had not gone to bed yet.

This is how it was in Argentina. Late nights were normal, so this situation did not bother anyone. Everyone ate quickly, half drunk and falling asleep. We made our way back to the car. We pilled six or seven people into the Mercedes and pointed it toward the vacation home. The sun was fully up now. The Brazilian sat on my lap so we could fit everyone. I sat up to look out the windshield to see where we were heading, and saw that crushed-up lines of white powder were being snorted off the dash.

I rocked my head back, exhausted. We left the city the same way we came in, fast. The girl on my lap had fallen asleep, along with everyone else in the back seat, and I fell asleep too. The air conditioning in the car felt amazing after being in the hot, humid nightclub on a long summer night. I passed out to the sound of tires screaming to hang onto the road as we took another turn far too fast for my liking.

13
Processing Cattle

The following day, I met up with my new employer. He was a veterinarian and a large landowner. All the neighbors called him Mack. He spoke perfect English, which made this whole situation possible. While my Spanish was really improving, I would still be lost without a bit of help. Most of his family had been educated in some of the best schools in Argentina. Most of these schools required that students be able to speak English.

The English taught in these schools was British English. The people who spoke English well in Argentina always had a British accent. Oddly enough, many of the people I spoke with wanted to spend as much time speaking English as possible. They wanted to pick up an American accent. The dislike for the British was strong in Argentina. This stemmed from the conflict on the Falkland Islands, where the British sunk a ship from Argentina in 1982. The battle killed men on both sides. Argentines are still quite bitter about it.

Mack was very welcoming and happy to have me working with him. We woke up early the next morning and drove out to his estancia. I met most of his hired men right away when I was there. I could not wait to ride and work with them. This was the first time I met gauchos who were not putting on a show for paying guests. This was the real deal. Immediately, they had me on horseback. A bull had escaped. Someone needed to run it down and bring it back to the working facilities. The boss pointed to a horse and said, "Get on that one and bring the bull back."

The saddle and equipment differed from what I had used farther south in Argentina. It was the same basic concept, but it all just felt different. Medicine bottles hung from the sides, and the saddle was greasy like they had used it for work for generations. The horse handled much differently. I could tell it was a cattle-working horse.

The most peculiar thing was that one stirrup was much longer than the other. The stirrup irons were tiny; I couldn't fit my boots in the stirrups. The bull was getting away, and I did not have time to do much more than kick the horse up into a trot to catch up. The saddle had no edges or hard pieces I could grab, so I did my best to pretend I was a little kid again, riding bareback. When I was ten or eleven, I saved up money to buy my own horse. I could not afford a saddle at first, so I would help do work bareback until I could buy my first saddle. I was comfortable riding bareback so this strange new saddle did not throw me off very much.

I immediately realized there was another odd thing about this saddle. It was so loose it was just bouncing all over, barely secured to the horse. I would find out later that most gauchos ride with very loose saddles unless they are roping something. The men didn't want to get off their horses and re-tighten them many times per day.

I caught up with the bull. My saddle had moved from one side to the other, but I did not have time to fix it. Mack raised some of the best Brahman bulls in the entire country; he often brought home ribbons from the livestock shows. I had learned a lot about Brahman cattle while in Australia. Two of the main points were that they can get mean in a hurry, and many of them are wild and will outrun your horse.

I was doing my best to make the correct choices easy for this bull so I did not have to fight him. The kicker was that since these cattle were also very well-bred, each one was worth a pile of money by itself. I was afraid I would make a mistake that could injure the bull. I got the bull back in the corral where he lived. I was just a disheveled mess of a foreign cowboy, way too far from home, but the hired men all smiled at me like I may have slightly impressed them.

There were more Brahman cattle in this part of the country compared to the southern regions. Brahman cattle are much better-suited for the hot weather in northern Argentina than a European breed such as Angus. Brahman cattle have big ears and skin hanging low from their necks. Both attributes offer better heat dissipation. Extra skin and ears allow more surface area to release heat when necessary. Unfortunately, this also makes these cattle perform poorly in cold climates.

Brahman cattle also have large humps on their shoulders. The hump is like that of a camel hump. The fat in the hump stores water, making them more efficient in hot and dry conditions. The Brahman breed and others like it originated in India and the surrounding areas. They have now spread across hot environments around the globe.

We had to brand some cattle that day, cattle too large to rope and brand like we would do with smaller ones. The facilities they used to work the cattle were one of the most impressive things I have seen in all of my travels. Everything was constructed using technology from a hundred years ago or more. Except for a few chains and bolts, they had made every piece of the chute from wood.

It took four men to run the chute to brand the cattle, and they did it with incredible precision. The animal would walk, or sometimes run, into the chute, and one man had the job of catching the animal around the neck. He pulled a long wooden lever that closed two boards around the animal's neck. It did not pinch the animal, but it was small enough that the body could not go forward, and the head could not pull back.

The same man then pulled a lever that held the animal's head in place to keep it from hurting itself or a person. Another man would then notch the animal's ear to show ownership. He would also burn the area where the animal's horns would grow. This process killed the tissue in that area, preventing the animal from growing horns. Horns could be dangerous for humans, horses, and other cows.

The man handling all the levers would have to sit on them or do some crazy yoga move to hold them all in place. Each lever had to be held because they could not lock in place. Letting go of one of these levers could be extremely dangerous for the animal and the people working around it.

There were two more levers at the back of the chute, usually run by a young kid, maybe thirteen or fourteen. First, he would slam a lever down behind the animal walking into the chute. This was to prevent the animals behind it from trying to come into the chute. He would then pull the second lever, which put pressure on the front side of the animal's hind leg. This prevented the animal from kicking, which helped avoid injury to the animal's foot from a kick to the heavy wooden chute. It also helped keep the people safe from rogue kicks that could slip between the slats. The whole process took only seconds: a man and a straining young boy pushed down and held all the levers in place. The man on the head did his work, and another man put a brand on the back-left hip of the animal. It was incredible how efficiently these men could process cattle with only these wooden chutes.

In addition to the owner's brand, each animal also got a four- or five-digit number brand, which was its identification. They did not use ear tags. This number was to track which animal was which for their purebred breeding program. On this estancia, they raised purebred Brahman, Brahman Black Angus cross, and Brahman Hereford cross. Each breeding program had its own numbering. The cattle on this estancia numbered well into the thousands, maybe even tens of thousands.

At one point, we ran an animal in the chute that had an infection on its face. The men noticed right away and knew exactly what to do. First, they made a small incision into the infection to allow drainage. Next, the men took a cup of iodine or a similar substance and poured it on the external part of the wound. Last, they needed to give the animal a shot of antibiotics to assist with fighting any potential internal infection.

The owner, being a veterinarian, knew exactly which bottle to grab off the shelf. He handed it to his cow boss and walked away. The man gave me a look of confusion, as if he wanted to ask me a question. He was afraid I would misunderstand it or cause a scene, so he waved to the little boy who had been running the levers on the back of the chute. This was his son. He asked his son the question he must have wanted to ask me. I understood it clearly, and I was shocked. He asked his son, "How much do I need to give this animal?" This puzzled me at first. Why would he not just read it himself?

I later learned that none of the men I worked with could read or write much more than their name. None of them had ever attended any sort of school. They were too poor to attend school growing up, because they needed to work instead. Most of the men were second or third generation on this estancia. They had grown up here, started work at a young age, and never left.

Most of these men owned a horse or two, usually a saddle, and that was it. They lived in simple houses built for them by the landowner. They worked almost every day, and that was life. My mind was blown. It was my first time meeting or seeing an adult who did not know how to read or write.

14
My First Saddle

The next day we went and caught a horse that would be my mount for my time there. The horse was a mare that Mack had trained years ago, but now had little time to ride. He told me his workers would sort it out for me, and then he left. He still had many other things to do besides babysitting me. The men led the horse up to me and told me they would "make me a saddle" so I could ride.

The gauchos' saddles had many parts, unlike English or Western saddles. First, we needed to get a pad, a layer to protect the horse's back against rubbing on long days of riding. They took an old couch cushion and cut the Styrofoam out of it to make a pad. Next, we needed the main saddle part. They had an old one with a bit of mold on it gathering dust in the shed, but that would do. The saddle consisted of little more than some rolls of leather stuffed with wool to give them some shape.

Then, the saddling got a little crazier. They laid a thick piece of leather with metal rings on both ends on top of the saddle. Hanging from each of those rings was a long piece of leather called a latigo, which was used to tie the saddle on tight. Usually on a Western saddle, a latigo would be attached to a ring sewn on to the main part of the saddle. To see it attached to a loose piece of leather draped on of the saddle was strange.

Finally, on the bottom they needed a cinch, the piece that would connect the two latigos from either side under the horse's belly. Usually, they were made of many strings woven together that perfectly fit the horse's underside. They sourced an old cinch from the barn. Then, a sheep hide was laid on top of the saddle. A saddle is quite soft; I never understood the need to put sheep hide on top as well. However, it was still useful because it added padding and would prevent me from sweating and sticking to the saddle.

The hide needed to be tied on with another set of similar latigos. We could not find another set, so they pulled out a piece of rope with the plan to tie it all together. Instead, one of the older gauchos jumped down and let me borrow the piece off his saddle. He disappeared behind some old rusty tractors, and came back with a ratchet strap, the kind used to secure loads on trucks. He threw it around his horse's belly and clicked it tighter until it felt right to him. Now, I had an entire saddle but no way to steer the horse.

We found an old bit hanging in the barn that had accumulated a good amount of rust. One man banged it on a rock to remove most of the rust and then used sandpaper to smooth it. Unfortunately, we could not find a rope strong enough to make a headstall, the part that holds the bit in the mouth. We also needed something to make the reins so I could pull to steer. A short man named Juan pulled out his huge, sharp knife and disappeared around the corner with an idea. He came back quickly with a long piece of an electrical extension cord.

With some quick measuring and cutting with the huge knives, they had all the pieces to make my horse rideable. Juan separated the three strands of wire that made up the larger extension cord. Without even looking, he began braiding the three wires together to attach them where they needed to go. Finally, thanks to these clever men, I had my horse, my saddle, and my reins. Some people might have been a little scared to ride using a setup like that, but it thoroughly impressed me. These men relied on nobody. There was no quick drive to the store to get a new piece. They made what they needed.

I swung up on my horse, the happiest man in Argentina. I was finally where I had dreamed of being in this faraway place. The men gave everything one last look, and we headed out to find the cattle we would work with that day. It was still dark outside. The gauchos chose to wake up at three or four in the morning to beat the heat. We rode along in formation, trotting side by side for nearly half an hour. Finally, we reached the paddock where we would gather, and the sun peeked over the horizon.

We opened the gate and split up into teams to get more done. We had two men named Juan on our crew, and I went with the Juans. When the sun gave the green landscape its first bit of gray light, I realized we were in heavy fog. The fog was so thick that I could easily lose the Juans, and I would not know my way back.

We picked up our gait, and galloped blindly through the fog. Both men were born and raised here. They did not need to see to know where to go. I stayed close behind as the hooves pounded and destroyed the wet green ground beneath them. I caught a few large hoof-shaped pieces of mud to the face, but that was better than getting left behind.

I suddenly noticed a strange but familiar sensation as the morning air passed over my face. The air temperature was correlated with the density of the fog. We would hit tiny pockets of warm air, and in those pockets, the fog was thinner, and I could see farther for a split second. This was a feeling that I had grown up with while riding motorcycles and horses. I wondered how many people had never felt this. My cheeks became ultra-sensitive. It pulsed quick chills down my spine when I went from a warm pocket of air to a cold one.

Only a person moving at high speed through the open air would ever experience this. I felt bad for everyone sitting in front of a screen that day. Heck, I felt bad for anyone that was not me at that moment. On a warm morning in Argentina, I was galloping through the clouds. I was holding onto my extension cord reins, taking dirt clods to the face, and I wanted to be nowhere else.

15
Muddy Toes

I would wake up every day around four-thirty or five. The gauchos would just be arriving then, but some of them had to ride an hour to work just to get there. They probably woke up at three to catch horses and saddle, make lunch and mate, and then ride to the main estancia. I felt lucky that I got to live right where I worked.

They gave me the royal treatment while I was there. I had the big house all to myself. The house had six bedrooms, maybe more. Beautiful swooping archways and antique furniture surrounded me. The family had lived in this house for generations before they had to buy a house in town for financial reasons. Now the house remained just for guests and special occasions. I even had my own maid at the house while I was there. I rarely saw her, but I could not leave a dirty sock on the floor without her cleaning everything. She would make me lunch almost every day. She usually made something called "Milanese." Milanese is a thin-cut steak, breaded and fried.

The house even had Wi-Fi. It was slow and did not work well, but it worked, which was enough for me; I could use the app on my phone to learn Spanish. During the days I stayed in this massive house, I believe I learned more Spanish than at any other time in my life. I was up early, working all day most days, but I was alone when at the house and could spend the evenings learning and reading.

I had a book called A Falcon Flies, which a lady at the first estancia I worked at had given to me. The book was about a brother and sister's adventure through Africa, which made me decide I would go to Africa one day. When I wasn't reading, I spent hours learning Spanish until my brain hurt. The next day, I would try to string as many words together as possible and use what I had learned.

Sorting cattle is usually a fast-paced event, and even with the best stockmanship skills, you need to decide and communicate quickly. I believe this taught me the most about being able to think in another language. Overcoming that was a massive hurdle. I needed to be able to ask, "Which one?" "What kind of medicine?" "What's wrong?" "what are we doing?" "why?" "How many?" "What number?" Then I needed to understand their response and react within seconds. If not, I would miss my opportunity and make a mistake. It was fantastic for my Spanish.

After about a week of this, I started counting silently in Spanish. Then, I began talking to myself in Spanish. Eventually, I told the gauchos some of the bull-catching stories from Australia. They understood most of it by the looks on their faces. I would ramble for a few minutes before I realized I had been speaking Spanish the whole time.

One of the first mornings at the estancia, I was up drinking mate, waiting for the men to arrive. I watched the sun spill over the horizon across the meadows full of tall green grass. Dew dripped off everything. It was already hot and humid. The soil in this part of Argentina was red clay, and it stuck to everything. First, I heard the clang of an iron gate near the cattle yards. Then I heard the screaming of old hinges begging for oil, and I looked to see who was coming. It was three children, aged maybe five to seven, all riding one horse bareback. They all wore small backpacks and carried plastic bags with their tiny lunches.

I watched as one child jumped off and opened the gate. The other two guided the old horse through the gate and closed it behind them. The child on the ground then reached her hands up and grabbed one hand of each kid sitting on the horse's back. She took her bare, muddy foot and swung it up as high as she could, barely high enough to reach the horse's belly. She dug her muddy little toes into the visible ribs of the ancient horse. Finally, she pulled hard on the other two children's hands, heaving herself up onto the horse's back .

All this time, I was watching in amazement and thinking of how self-reliant these kids were. The horse they rode must have been older than me. It had a deep, sagging swayback but it plugged along flawlessly, escorting the kids to school. I remember watching that, and the words to a Wade Montgomery song came to mind, "I got a weakness for them old swayback horses." The image of that girl climbing back on that horse will forever remain in my soul.The girls' siblings hauled her back up onto the horse, and they rode off. I thought about how much I hated school at that age, how much most American kids hate school at that age. We all listened to stories of our ancestors riding to school uphill both ways in a blizzard way back when. Well, here I was witnessing it in real life.

I mentioned to one of the gauchos that I had seen the kids riding by that morning. They were his kids. He told me none of the adults I worked with had ever gone to school. Most had never even seen the inside of one. I had only witnessed the kids opening one gate, but he said there were many more gates along the route. It took them more than an hour to get to school. They rode rain or shine; there was no other option.

They only had the opportunity to go to school because Mack and a few other surrounding landowners built a school and paid a teacher. They could go to school up to fourth grade. After fourth grade, they would have to go to town for school, but usually they stopped at that point. This was why the cow boss had needed to ask his son to read the medicine bottle to him a few days earlier. His son could read and write, and he was damn proud of that.

16. Gauchos of the North

We rode nearly every day for most of the day, so I got to spend a lot of time with the guys and learn about them. Many of them had married at a young age. Most of them had their first kids when they were around eighteen, which was common in poor parts of the country. I could not imagine having a kid at that age. I thought the men married young until I found out most of their wives were three to six years younger than them. Some of their wives were even younger than that.

The girls married very early; some of them got married at fourteen, which was just shocking to me. They told me it was because girls rarely brought income to the household. Parents were happy to marry them off as soon as possible; then it would be their new husband's responsibility to support them.

This was something that the women seemed to just nod their heads and agree with. This certainly was not the case in the cities. Many cities were very modern and had adopted a modern outlook on women's equality. However in the rural places, where the written word did not go far, tradition was king. Women seemed to hold the same rank as a man's favorite horse. They were extremely useful and much loved, but they were not a man. I had heard about these things before in books about the history of the United States, and I had seen it in documentaries of faraway places. However, this was the first time I had witnessed it in person. It was eye-opening.

When I traveled to northern Argentina, I noticed a change of attire in the gauchos. They were more colorful and expressive than the southerners in every way they dressed. Their hats were a little bigger, their shoes were in bright reds and blues, and they had a strong sense of pride in their appearance. In addition, many made their own equipment for daily use. Most of the tack was handmade, the whips were hand-braided, and many of the clothes worn by the gauchos had been made by their wives.

The shoe of choice was the "alpargata," which was a shoe made of thin cloth with a very thin sole. They were exactly the kind of shoe that TOMS sells. Some men wore name-brand TOMS. Others wore knock-offs. These shoes offered no protection for the foot, and the ankle was totally exposed. They used denim or any material they could find to protect the ankle and keep dirt from getting in the shoe. They wrapped their legs from just below the ankle all the way up to the knee. These wraps resembled an English rider's half chaps. They would make them out of the brightest colors they could find. They were beautiful but faded quickly from the sun and dirt and lost their zest.

The men of Corrientes were also famous for their knives. They kept exceptional knives that were always razor sharp. They were quick to draw them in a fight, even more so than gauchos in other regions. People warned me multiple times to stay away from "gaucho bars." I wondered if they knew I worked with gauchos all day. But nevertheless, I took their advice and stayed away.

I was always seen as the rich gringo because I could read, write, and speak English. Unfortunately, gauchos had been taught from a young age that outside of work, they did not mix with people like me. To walk into a room full of drunk people with big sharp knives probably would not have been in my best interest after all.

17
Wild Cow Roping

One thing I could not wait to learn was how to use the rope
that the gauchos used. They used a lasso like the original
American cowboys and Mexican vaqueros. They braided the
rope out of rawhide. The more strands the braid had, the stronger
and softer the lasso was. Since I was the new guy, they gave me
the lowest-quality lasso, the kind every good gaucho learns to
use before moving up to the more high-quality ones. I did not
use mine yet because I wanted to watch their technique first.

One day, we rode to a large pasture early and started bringing
in all the cattle. We were going to give them vaccinations and dip
them that day. When we got close to the pasture gate, one old
cow looked at her calf, who was happily in line with everyone
else. This cow decided he was on his own, turned her head, and
bolted straight back at us.

We all tried to stop her. She slipped through the line of horses;
whips cracked. Different men tried physically turning her with

their horses to no avail. She could not be stopped. Finally, one man went with the cow at a run so we would not lose her. The boss yelled at me and told me to go help the man. Laughing, the boss said something, which I believe was, "Careful, that cow might give you a real Argentina welcome."

I ran across the pasture after the man and cow, trying to keep up. We hit the backside of the pasture, and she busted through the fence and kept going, so we did too. Soon we were in the middle of a group of about twenty or thirty yearlings trying to keep an eye on the cow. We circled the yearlings, trying to keep them in one group. We could not find the cow because they were all white Brahman, and she was smaller than most other cows. It was hard to keep track of her.

We hoped to keep the group together until reinforcements arrived to help get her out. At this point, a good stockman would have called it quits. We should have put her calf out in the pasture and waited for her to return and get it; now it was a matter of pride. Opportunities for fun like this rarely arose.

Finally, a few more guys arrived, and we planned together. We would ride into the herd, and they would scatter. Whoever saw the cow first would stick to her like glue, and the other three men would come as quick as possible to help. They told me to get out my rope. One man jumped off and showed me how to attach it to my saddle. We pulled the latigos tight and tied the tail of the lasso to the latigos.

The latigos were the only place to use as an anchor since the saddle had no actual structure. Basically, the lasso was tied to a strap wrapped tight around my saddle and the horse's belly, and I sat on top, wondering if a good tug would yank the saddle to the side. The other men built loops so they would be ready to throw at a moment's notice. First, they made one loop about as big as a man to throw around the cow's neck. Then, in both their left and right hands, they made smaller coils of rope they would release when they threw. It gave them the ability to throw the loop further.

We rode into the herd slowly, ready to throw a loop as soon as we saw her, and the cattle shuffled. The shuffling turned into a full-blown bovine explosion. Cattle began scattering in every direction. I hoped my horse knew well enough to maneuver itself out of the way of any stampeding creatures. I began looking for the only discernible physical feature that would set this cow apart from the other animals in the bunch, her udder. I bent down, trying to see the underside of every animal squirting out of the bunch. Then I heard the men yell. She had tried to escape from the other side of the group. One man was right on her tail, and we needed to get there to help fast.

None of these men rode with tight cinches other than in moments like this. One man had decided before this happened that he would help me tighten up. They did not trust me to do anything alone. When I had tried to tie a T-knot in my latigos to keep them from loosening, they told me, "No, no, no." Tying a T-knot was how I had grown up using my Western saddles. They insisted that I pull it tight and then sit on the end of the latigo.

However, no matter how much you tightened a latigo, it would invariably loosen with time. Well, what a time for mine to loosen, right in the middle of a wild cow chase. I was still in the frenzy of the main herd. I ran, trying to catch up with hardened gauchos, swinging a loop over my head that I had never swung.

I caught them, my horse clicked it in high gear, and we were a part of the madness in no time. When I arrived, two men had already thrown their loops and missed, so they were trying to rebuild their loops. The last man with a rope ready was in position to throw, and he missed. It was my turn. I rode up into position. My heart racing, I swung the loop over my head effort-lessly. I remembered all the times as a kid I had roped a fake calf head to practice.

Just as I was about to throw, the cow ducked left. No problem, I thought. Out of pure instinct, I pulled slightly left on my reins, and as soon as I saw the cow, I let the loop sail. The loop was beautiful, perfectly cast right at her head. I already saw myself being the hero of the day when I suddenly tasted dirt and felt my face hitting the soft ground. In my haste to be the hero, I had not noticed my saddle becoming very loose. As soon as I put too much pressure on one stirrup, I was off the side and on the ground, a fool. I had completely missed it.

The other men looked back and saw me stand, so they continued the chase. I was completely uninjured. It had not been a hard fall, maybe harder on my ego than anything. I caught my horse and got my things in order. I collected all my shit that I had just made a yard sale out of across the red mud and jumped back on my horse. By the time I reached the men again, they had her caught. I had watched it happen from a distance as I raced to catch up.

I was happy I didn't catch the cow first, because I would not have known what to do next. I had never roped something with my lasso tied to my latigos like this. Ultimately, all three men threw their ropes on her, guided her to the correct paddock, and then back to the corral. I returned covered in red clay mud, fully intact. I was not even slightly disappointed by my first experience with an Argentinian rope.

18
Wildlife

In the few weeks I stayed at Mack's estancia, I learned incredible amounts of horse and cattle wisdom from the men I worked with. Many of the diseases we saw there differed greatly from the health issues I was used to in the mountains back home. In the northern parts of Argentina, there were a mix of tropical and grassland diseases. One of the worst ones I saw was screwworms. Watching them be removed almost made me heave. A parasitic worm would burrow under the cattle's skin, often in the neck. All you could see were bumps all over the cattle's neck. Sometimes the bumps oozed infection.

The only way we could get the worms out was to hold the animal down. We did this in the pasture using lassos or in the chute at the working facilities. They would squeeze the small parasites out like popping a zit. That was the part that always made me sick. Next, they would pour some disinfectant on the wound so it could heal properly. They warned me not to get the worms on me because they might burrow into me. I was also quite cautious about mosquitos and insects. Nearby people contracted the Zika virus and Dengue because of mosquito bites.

I saw many amazing wild animals as well. We often saw "Dorando," what the locals called the native large, flightless birds. The more official name was "Rhea." They are a cousin of the ostrich, and we would sometimes run after them to see if we could rope them. We never came close. Burrowing owls were everywhere too. These owls lived in burrows that they shared with wild chinchillas.

Maybe the coolest animal I got the opportunity to see was the "Iguazoro," as the locals called it. The English name for it is Maned Wolf. They are one of the strangest canids I have ever seen. They look like they have the head of a red fox with the body of a wolf and exceptionally long legs. The only one I saw was at night. They are beautiful but almost nightmarish-looking creatures. I was told they are rare; my boss was with me and had only seen a handful in his lifetime. I felt lucky to witness that.

After three weeks, my time was over at this estancia. I was headed off to stay with another veterinarian, who was a friend of my boss. He told me the next estancia was even bigger. They owned all Hereford cattle, and it was mainly swampland. I jumped on a bus and rode for a few hours to meet my contact. He spoke little English but told me someone was coming to give me a ride. When the man arrived, he was very talkative but spoke no English. We rode for an hour out into the middle of nowhere.

When we came to a stop, the driver told me the long driveway was too much for his car to handle, and I would have to wait here to get picked up. I sat down on my bags in the middle of nowhere, at a crossroads, waiting for whoever was bringing me the rest of the way. Finally, I heard a tractor. I did not know who was picking me up, so I assumed it was this guy. He jumped down to greet me, shook my hand, and took one of my bags. He tied it to the hood of his big, open-cab tractor, then took my backpack and did the same thing.

I jumped up and rode on the loud, bouncy tractor fender all the way to the estancia. It took us forty-five minutes or more by tractor to get there, bouncing through deep mudholes and over rocks. Sitting on the fender made my back sore, and I understood why you would not attempt to get a car down this road. The driver merely suggested a direction for the tractor to go; it rolled over the path of least resistance. The tie rod ends of the tractor's steering were completely worn out, so the tractor's wheels wobbled back and forth.

I was really in the middle of nowhere now. From the nearest bus stop, it had taken an hour by car and then forty-five minutes by tractor just to get to the main buildings. If I needed to get out of here, it would be hard. Plus, nobody had told me how I was getting out. The people who lined this job up for me said they would get me when I was ready to go to the next place, but there were no more details than that.

Again, they gave me a nice, big guest house with a maid who cooked for me. These big houses just sat vacant. The families had all left the houses and moved to town for easier living. This house sat right on the edge of a massive swamp where I could see the world's largest rodent, the capybara. They swam in the swamp, trying to avoid snapping caiman and snakes. I had never lived near a swamp before. It was both beautiful and haunting to me at the same time.

The nighttime was the worst. The sounds that came out of there did not let me rest well. The bugs were everywhere, so I tried to cover up as much as possible. It was hard to avoid them because the house did not have screens, and it got extremely hot and humid at night. You had to leave the windows open to keep cool. I did not own a mosquito net to sleep under. The best thing

I could do was put an old fan at full speed. I hoped the blowing air would keep the mosquitos off me, and it mostly did as I slept.

The first night the maid came to fetch me for dinner. She had just finished cooking my meal of meat, soup, and potatoes. It was delicious. After traveling all day, I was starving. I sat alone in the big house and ate my meal, the maid checking in on me all the time to ensure I needed nothing. I watched as a huge black spider crawled from behind an old picture frame on the wall. He was bigger than the palm of my hand. At this point, nothing really startled me, at least if it kept its distance. He respected me, and so I did the same to him. I ate my dinner, and he sat on the wall not far away, watching.

As I was eating dinner, the shrills and screams of the swamp nightlife crept in through the windows and into my ears for the first time. The air was full of monkeys screaming and horrible noises from other wildlife I knew nothing about. One last time, the maid poked her head in. She told me I needed to speak with the head stockman when I was done. He was waiting outside. He introduced himself, shook my hand, and told me to be up and ready to work at six. "No problem," I told him. That was it. He turned and walked away, and I returned to my room to sleep. I knew I would not sleep much between the noise, the mosquitos, and the excitement for what tomorrow would bring.

19
Swamp Herefords

The next morning, I was up before the sun. I checked around for spiders and then slipped on my shoes. I filled my reused water bottle with some water that had a brown tint and went outside to meet the new guys. There wasn't anyone in sight; I was a little early. I thought maybe I had misunderstood the time or place I was supposed to meet them. I sat there in the dark silence, smoking a cigarette and letting my mind wander. My brain was tired.

While I was learning Spanish, I spent my entire day trying to rev my brain up to full power just to understand a simple conversation. I realized that language was something I had always taken for granted until this trip. I snapped back to reality when I heard the thunder of a large horse herd approaching. It was dark. I could only guess which direction they were coming from by the sound of their hooves. I turned to see one man with a cigarette hanging out of his mouth crack his whip and turn the horses into the corral.

The remuda contained at least fifty horses, all with neatly-trimmed manes that were all slightly different from one another. There is an entire code related to how a horse's mane is cut, which vaqueros use as well. Different cuts in different places tell the rider what kind of training the horse has had and if it is wild or gentle. Finally, the men appeared out of the darkness and grabbed halters to catch horses. The boss handed me one as well.

When the horses came in, they were expected to show respect. The horses turned and faced the men, standing perfectly straight in a line. If one turned its butt to the men, they would take a rope and smack it in the butt until it turned and stood in line like the rest of the horses. They stood in front of a three-hundred-year-old stone wall. Slaves had built the wall. The horses stood like a regiment of soldiers waiting for orders.

The men walked up and grabbed the horses they wanted. The boss turned to me and told me to grab a big gray gelding toward the middle of the line. I walked up, slipped his headgear on, and led him away. When they were finished, the horses received a whistle signaling they could leave the corral. In perfect order, they fell out of line and stormed back into the pasture. I was so impressed; this was the first time I had seen horses behave this way. No doubt years of training were involved, as well as the older horses teaching the younger ones.

That day we would vaccinate a large bunch of cattle, but we had to gather the cattle up first. This ranch raised only purebred Hereford cattle, and they raised thousands of them. It was not the ideal breed of cattle to raise in these tropical conditions, and the cattle reflected that. Many of them were skinny, but they were surviving. We rode out at a trot just as the sun came up.

The men asked me the normal questions. Where I was from? Did I have kids? Was I married? Did my parents own a ranch back home? It passed the time as we rode far into the swamp looking for cattle. Finally, we reached the gate. The boss drew a picture of the pasture in the dirt with a stick. He showed each of us where to ride, so that we would not miss any cattle, and we all rode out.

It was an easy gathering. The cattle were very tame and knew exactly what to do. Just seeing a horse and rider signaled to them that it was time to go to the corrals. We had an entire pack of dogs with us, and none of them helped with the livestock. They were just companions. Mutts of every shape and color ran along behind our horses. I heard the dogs start to snarl and fight. I looked up to see them all piled on top of an animal, trying to rip it to shreds. There were so many dogs in the pile that I could not even see what they were attacking. I was terrified. I thought it was a young calf or one of their own that they were trying to kill.

One man rode up to them, looked down calmly, and then rode away. I thought either they were killing something that was allowed, or whatever they were killing was too far gone to be helped. I rode up as well to get a better look. The creature being attacked was a capybara, the massive rodent-like pig things that lived in the swamps. The animal was doing far better in the fight than I thought. It was throwing off the shaggy little mutts left and right. One man said, "If the dogs get hurt, it's their fault for picking that fight." So, we all just rode on and let them fight it out. About five minutes later, all the dogs showed up looking relatively unharmed. None of the mutts had a piece of capybara in their mouths, so it must have been a draw.

We spent all morning gathering the mothers and babies. Then, finally, we got them back to the dusty, old handmade corrals. We put them in a corral and let them settle down as we ate lunch. First, we planned to give them a vaccination, and then they would be jumped into a dip. A dip is a long, deep pool, about as wide as a cow, with concrete sides. The cow jumps into this pool and swims across. Every bit of the cow gets wet in the process. The purpose of getting wet is to get the cow covered in the chemicals that we dump into the water. They are not harmful to the cattle but kill ticks, worms, and insects that feed off the cattle. It is basically a pool full of bug spray.

After we ate lunch, everyone smoked a cigarette. The men showed me that they chewed tobacco, and I showed them how Americans chew tobacco by stuffing it in my lip. Then it was time to set up the dip. We added water to the pool and then two types of chemicals. These dip tanks are deadly for people because they still contain arsenic from many years ago. Arsenic was an old-school way to remove parasites from cattle. The men had me dump a white, powdery substance in the water. They told me to cover my eyes, nose, and mouth. If I breathed it in, it would make me sick. Then, they dumped in the liquid agent.

Just two gallons of the chemical were potent enough for the entire massive pool of water. They used sticks to mix it up. I was told I would push the cattle from the back. The experienced hands would do the work near the dip. If someone fell in and got their body covered in the chemicals and old traces of arsenic, it meant certain death. If a cow jumping in splashed the liquid in a person's eyes or mouth, they would be in awfully bad shape.

One of the older men told me he worked on an estancia once in another part of the country. When they drained the dip pool entirely for the first time in years, they found human bones at the bottom. I was happy to let them do the work around the dip, so I could stay back and deal with loading the cows. First, we had to separate the cows from their babies. We did this on horseback. The men would ride in and start removing the cows from the babies. It was simple work until only a few cows remained. They did not want to leave the middle of the bunch of bawling calves.

When one of the last cows was finally on the edge of the bunch, they used two horses in the rear to physically push her. They kept another horse to her side to keep her from turning, and then they made her run down the fence toward the gate as fast as possible. While running, she had no chance to even think about stopping. Three horses and one cow went barreling through the gate at top speed. They had to do this with about ten or twelve ornery old cows. It did not hurt the cows, horses, or men, but it certainly was not "low stress" like many producers strive for these days.

Once separated, I brought the cows up through the chute. Two men gave vaccinations to each cow, then two men pushed them up to jump in the dip. After the first cow splashed into the dip, there was a constant splashing of deadly chemicals for the next three hours. I could never find out what was in the chemicals that made them so bad. Nobody seemed to know what they were called, but the veterinarians provided them, so they must have been safe for cattle.

By the end of the day, I was exhausted. It was hard work. I was hot, sweaty, and covered in dust. The facility had no hot water, which was fine with me. All I wanted was a cold shower. By this time, it was February. All my friends back in the United States were complaining about the cold and staying inside. I was tan and lean from all the hard work outside. I did not have any desire to go home.

20
The Wild Mare

Growing up, I heard stories about how the horsemen of Argentina were among the best in the world. The Spanish arrived in the Americas with horses centuries ago, and the old Spanish methods can still be seen today all over the Americas. Ranging from Northern Canadian ranches to estancias in Southern Patagonia, many of the basic skills and principles of horsemanship are the same.

However, the small differences make it awfully hard to transition between riding styles, which I quickly found during my travels. Even within the country of Argentina, the horsemanship was quite different depending on the region. The men and horses in the south were bred for the mountains, short, stocky and sure footed. In the polo-obsessed estancias of the pampas in the middle, men rode tall, lean horses built for speed. The men riding the large footed heavy boned swamp horses of the north used very little pressure with their feet. This surprised me.

I think most expert horsemen agree that feet are essential to having incredible control of a horse. This was the first and only horse culture I encountered during my travels that used little to no foot pressure. It was obvious in the northern part of the country; they focused all on the mouth of the horse. Some men used severe bits in their horse's mouth to make him more sensitive. This was not how I was taught to do things, but I was just there to learn, not to teach.

I heard rumors of men riding with just their big toe in a ring instead of riding with their entire foot in a stirrup. I never saw someone do this. I heard men mention this type of riding when I was there, though, so it must have existed. The closest I saw were men riding barefoot in small metal stirrups. Getting your bare foot stuck would certainly hurt. However, these little stirrups were so small their entire foot could not go through, so I don't believe that was a risk.

One day, the boss said we were going to catch a renegade wild horse that roamed the property. She was born on the property and belonged to the owners, but she was so wild that they had never caught her. The wild mare was approaching seven years of age. A monstrous-looking black cloud loomed over us in the distance. We grabbed some ponchos in case it should rain, and we took off searching for this wild mare.

It had rained the night before, and the ground was slick and muddy. Both men and horses were covered in mud before the work even started. When we entered the pasture, they told me, "There is nothing in this pasture but the horse. So, if you see anything moving, it's probably her." In silence, we fanned out, straining our eyes for a glimpse of the wild, bay-colored mare.

A man saw her and alerted the rest of us with a series of short, sharp whistles. We all moved as quickly as possible toward her location. She popped her head up and looked right at us. We had been spotted. One man released a horrific war cry and began pounding through the short, heavy brush toward her. All the other men, including me, followed in pursuit. The mare exploded into a gallop.

Soon she was running down a two-track road, and we all fell into line behind her in pursuit. The maverick mare was the conductor at the front of the train. Like train cars, a string of rough-looking Bandito gauchos gave chase, along with one gringo trying to keep pace. At one point, we ran four horses abreast down the muddy road. Our ponchos flapped savagely in the wind as we galloped along.

The men had tried to be polite and give the gringo a horse that would not buck him off, and I appreciated that. However, my horse was older and slower than their young, fire-breathing mounts, so we had a terrible time keeping up. When we began falling behind, I felt the familiar pelting of hoof-shaped mud clumps. My horse took a few to the face, not injuring him but distracting him.

My horse was thinking about avoiding flying objects and not where to place his feet. He tripped, nearly tumbling ass over tea kettle into the wake of the gaucho locomotive. When we gained our footing, we had lost too much ground. My horse took me into the chest-high woody brush, hoping to find a shortcut to catch up.

I could see the men circling around. They would likely come right back toward us. We tried to move as quickly as possible through the brush to meet them. Unfortunately, my old nemesis snuck up on me again when I least needed it. As we hustled through the brush, we had to jump back and forth, left and right, to take the path of least resistance. Finally, my horse zigged, and I zagged. The incredibly loose cinch that held my saddle upright could not help me. I toppled off the side. I looked up to see my saddle on the side of the horse.

Twice now, loose cinches had gotten me off my horse. Twice I was told I could not tie them so they would stay tight. This was not how they rode. I suppose it was my poor riding both times. I put far too much weight on one side, which kept tipping me off. After this, I rode more like I was bareback, putting little pressure in my stirrups.

I sat fixing my saddle as I watched the men fly over the top of a hill, still chasing the bay mare. They flapped their ponchos to keep the mare running as fast as possible. What happened after that, I could not say. I was out of sight and earshot, but they caught her without me. The thunderstorm had arrived. I put on my slicker, mounted my horse, and hurried to catch them.

When I caught up, they had the mare in a halter, one man leading her and one behind to keep her moving. We trotted back in the downpouring tropical rain. It rained so heavily that it was hard to see where we were going. The wild bay mare trotted along, head high, still too proud to admit they had caught her. Her tail and mane were full of knots, and her eyes blazed as if you could see her spirit in them.

We returned to the barn. We took the mare to a stout wooden post and tied her there so she would not pull loose. We rode into the barn and unsaddled our horses. All the men and horses looked the same. With so much mud, you could hardly see the colors on them. The horses got a small helping of grain as a reward for a hard day's work, and we released them. They all walked out and just stood in the warm rain, letting it wash the sweat from their hides. When they felt clean, they trotted off to join the other horses in the herd.

The wild mare sat there, tied to that post in the rain. She pranced and pawed and pulled, but she could not get free. She was just beginning her training process. The boss called it a day. We sat around and listened to the radio on one man's ancient cell phone. It was a brick-shaped, off-brand Nokia. The service was poor this far out. He rolled up a piece of aluminum foil and stuck it in one of the phone's holes as an antenna. We listened to music for a while before changing into dry clothes and cooking dinner.

21
Hands

The wild horse we had captured now needed to be trained. They informed me they would do it gaucho style, the traditional way. In the dark the next morning, the horses rumbled in and lined up like normal. I grabbed my trusty old gray horse, and they all caught their mounts for the day. We led the ponies up to the barn and saddled up. We cleaned the mud off our gear as we went, which had accumulated during the previous day's circus.

The short man, who was the most talkative of the group, said the mare was his to break. He loved to break the wild ones. I stood by and waited for him to saddle the mare. I knew it would be a rodeo, and I would not miss it. He did not saddle the mare. He saddled a different horse, and we went about gathering and working cattle as normal. The mare sat there tied all day.

By now, the mare had settled down. She watched us work, and I imagined her wild and carefree life crumbling around her. She had not been given food or water since we caught her. After we finished working, I offered to give her some. They told me no. It was part of her training to be off feed and water. I thought it was not my place to say anything, so I did as I was told. The next day, we again followed the same procedure, catching horses. Again, the mare stood there tied to the post, going on her second day without food or water. We left as the sun came up to gather another large paddock of cattle.

When we reached the far end of the pasture, we came to a large stand of trees. The men said they were fruit trees, so we rode over to them to see if there was any ripe fruit. The trees sat next to the water's edge in the swamp, and it was clear animals had been living under them. I could see their prints in the mud. I stepped off my horse and walked up to the water's edge, where I gathered a few magnificent-looking shells. Some belonged to snails, others belonged to some kind of mussel. All the shells contained fascinating colors.

I jumped back on my horse and rode on with the other two men. They peered up in the trees with a slightly alarmed look on their faces. We had ridden deep into the trees now; they were very tall, and the edge of this tree line was like the tips of the jungle's fingers reaching out into the rangeland, trying to take hold. There was not enough moisture to sustain enormous trees like this away from the swamplands.

My pocket full of shells rattled as we rode, still looking around for fruit. The men became more concerned. Finally, they turned around and told me something I didn't quite understand. The only words I got out of it were "danger" and "hands." I thought to myself, "If there is something dangerous for my hands in here, I'm not touching anything but these reins. Then I can turn around and get out of this creepy jungle if I need to."

I scanned the trees to see what the men were so concerned about, but I saw nothing. When I looked back down the trail, both men were gone, but I could hear them talking farther up the path. I leaned forward and low, hugging my horse's neck to get through the brush when I heard the men ahead yelling, "HANDS! HANDS! RUN!" They came exploding past me in the other direction, signaling me to run like they were being chased.

I spun my horse around and headed for the daylight at the forest's edge. Just then, I saw what they were running from, and it all made sense. In Spanish, the words "hand" and "monkey" are very similar, and I had gotten them mixed up. The entire time, they were warning me about monkeys, not my hands. I saw a small bunch of them swinging toward me high in the canopy.

They were not big, scary monkeys; they were cute little fuzzy things that looked like pets. However, they knew something I did not, so I moved out. The monkeys caught up to my horse and me. We could not move quickly through the slippery mud, moss-covered logs, and vines. The monkeys were directly above us now.

I thought maybe they would bite me or something, but what they did was beyond my wildest imagination. It changed the way I look at monkeys to this day. I looked up to see a cute little monkey above me. He looked me dead in the eye and, with his little monkey hand, grabbed his little monkey dick and started trying to piss on me. I was appalled, and men's concern suddenly made sense to me. I tried to move as fast as my horse would go when I heard a splat right next to my horse, then another. I realized they were not only trying to piss on me, but they were now hurling shit balls at me.

My horse seemed unphased by the fact that we were now in the middle of a monkey piss hurricane. These were not hailstones falling from the sky: they were little monkey poos. We finally emerged into the grassland. We were safe from the little tree-dwelling shit-slingers. I stopped to look around. The two men had emerged not far from me on the edge of the forest, and they just laughed and laughed when they saw me. Apart from a few little piss dribbles, I had emerged unscathed from the affair. The men did not care. They had given the gringo a gaucho education.

When we returned to headquarters that day, the wild mare still stood at the post. She was not so wild now; she hung her head low. Horses, especially wild ones, can withstand more than people give them credit for. Now with two full days without food or water, this mare had finally lost her step. The men decided it was time to saddle her. Four men helped get the job done.

When the men approached, she immediately returned to her feral mindset, striking at them but missing every time. In her weakened state, she was not up for the fight. They had her saddled in less than a minute, and then one man put a blindfold over her eyes. They untied her from the pole, and another man on his horse grabbed her lead. The short man with a big mouth graciously jumped up on her back, and with her eyes covered, she hardly noticed.

When both men were ready, they pulled her blindfold off, and the rodeo began. She bucked, jumped, rolled in the air, snorted, and did what she could to remove the little man. As they said in the old cowboy song Zebra Dun, "He was grow'd there, just like the camel's hump." The mare decided it was time to bolt. She ran off, and the man on the broke horse went with her.

Both horses galloped down a wide path, and the man on the broke horse guided the wild horse so she would not hurt herself or the man on her back. They ran far out of sight, where the men had strategically placed a fresh horse and rider. When the two horses reached the fresh rider, they handed off the rope. The wild mare ran with the new, fresh horse until she got tired and gave up. They slipped a bit into her mouth and released her head. The little man turned her back toward the barn. As the sun blazed a fiery red sunset behind him, he rode her back into the corral where they had started.

It was not pretty; the horse had been blindsided and did not know what had hit her. Now, only a few minutes after the first human hand was ever laid on her, a man rode her around the corral, smiling like a villain. She would not buck or put up a fight; she was far too tired and weak now. She had much to learn. The men had conquered her wild spirit easily by taking a somewhat unfair advantage.

That night, she was unsaddled and rubbed down. She was bathed and fed a large helping of hay and oats fit for a horse queen. She was given water as well. They wanted to replenish everything they had taken out of her. The next day, the little man saddled her up without help. He swung up on her back and rode her all day, working cattle without an issue. It was certainly an old-school, rough way of doing things, but they impressed me.

22
Sickness

I had now been at this farm for nearly two weeks. I was having a blast. Every single day, I got to experience the real deal. I had come so far from the hospitality experience I first walked into when arriving in Argentina. The mosquitos even seemed to stop bothering me. My Spanish was improving by leaps and bounds, and I had found what I wanted.

I had a fuzzy feeling in my stomach, and suddenly I started noticing that I got tired easily while working. From one day to the next, it was getting progressively worse. At first, I thought maybe it was something I had eaten, or maybe just poor nutrition. When I felt alone and uncomfortable, I usually smoked a ridiculous number of cigarettes. Maybe that had caused it.

I initially brushed it off and thought it would improve with time, but it only worsened. It then attacked my morale. I no longer wanted to think all day, every day, to communicate. I was tired of being broke and stressed about money. Oddly enough, the risks I took around horses and cattle no longer seemed worth it. I decided if this got any worse, it might be ugly, and there could be something seriously wrong.

I decided to leave the estancia and try to get back to Buenos Aires and fly back to the United States. This was no easy task, given how remote I was. I had to tell the maid to please call someone to give me a ride to the closest bus station because I was not feeling well. She somehow misunderstood me and thought something was wrong with my family. I tried to explain it differently to her, but it was no use. I just needed to leave.

I again took the long, bouncy ride down the driveway on the tractor with wheels that turned wherever they pleased. On the ride in, I had been full of excitement, with an adventurous spirit, and ready to work. Now, I was worried. I felt awful for having to leave them on short notice. I was so nauseous I could barely keep the food in my stomach as we bounced along. When I got to the dirt road, the man on the tractor dropped me off and left. I sat there again, with no idea who I was waiting for. They had said I was getting a ride.

A man in an incredibly old ford pickup pulled up. He told me he was taking me to town. I threw my bags in the back, and we drove for a few silent hours into town. I paid the man way more than I expected, but I felt so bad at this point that I did not care. He took me to the bus station, where I jumped on an overnight bus heading south.

When I arrived at my stop the next morning, I went to buy my connecting bus ticket to Buenos Aires. It was Sunday, and a few buses were running, but unfortunately there were no more tickets to Buenos Aires until the next day. This was not a tourist town, and there were no hotels or hostels to stay in. I knew I would end up on a park bench if I did not figure something out.

I found one woman working who spoke some English. I explained my problem to her, and she had no answers. She said all she could do was put me on a bus to Cordoba. This was not ideal, but my friends were still living there while attending university. So, I thought I would visit them for a few days rather than sleep on the street in whatever town I was in.

When I showed up at their apartment complex, they were surprised but happy to see me. They had moved one door over to a slightly bigger apartment, one that had a real couch for me to sleep on. I spent a few days there, simply happy to be inside, and my condition worsened. I had booked a flight home from Buenos Aires, and now I needed to get another bus ticket. I walked around the city until I found a place that sold tickets. I bought one for the next morning. When I left my friends, they could see in my face that I was declining. I was pale and weak. They wished me luck, and we said goodbye. I slung my heavy pack onto my back and began the long walk to the bus station. I climbed on my last bus, bound for Buenos Aires.

I had booked a night or two at a hostel in Buenos Aires to tide me over until my flight. I knew my Spanish had improved because I checked in totally in Spanish. The hostel workers spoke English, but I did not need it. I did not realize it until I turned to follow an employee up to my room. She said, "Oh, you're American? We could have checked you in speaking English. I didn't know." I spent my days there lying in my bunk bed, trying not to vomit. I would rush to the toilet a few times a day and puke up blood.

I would cook simple little meals for myself, food I thought I could keep down, but it never worked. There were so many interesting people buzzing around that hostel that I wanted to talk with. They were from all over the world, and all of them spoke English. I just wanted to sit and talk with someone, but I was so sick I could not.

My head pounded with a headache, and I just kept counting down the hours until I flew home. Finally, it was time to call my taxi and go to the airport. I shuffled outside and jumped in.

When I got to the airport, my vision was getting blurry. I felt the worst I had felt yet. I wondered if I should talk to someone about getting medical help at the airport. I decided against it and boarded my flight.

It was the longest flight of my life; I dry heaved the entire time. I could not sleep. I felt so sick. When I closed my eyes, I just felt sicker. I went to the bathroom probably twenty times to vomit, but I ended up just dry heaving up more blood. When I finally landed at home, I secured a doctor's appointment for the next day. Just being back on home turf gave me a bit of comfort.

When I saw the doctor the next day and told her all that had happened, she was alarmed. She began a barrage of tests on me to eliminate possibilities. When she came back with some results, it was not good. She wanted me to go get my guts scanned. When I did, the results shocked her. She had to look at the result a few times, and even then, she did not completely understand what was happening.

My liver had shut down. It was in the first of three stages of failure; the damage was not permanent yet. She then needed to run more tests to understand why. She said I had the liver of a forty-year-old alcoholic. It made no sense. I drank and partied more than the average person, which was partly the reason, but my liver should not have been this damaged. She thought I might have hepatitis A.

The test came back negative. The doctor then thought I had contracted the Zika virus from mosquitoes. She said we would never know unless I wanted to do a lot of expensive tests. If it was Zika, I would not be contagious to anyone, and all I could do was wait it out and take it easy so my liver could heal. This scare was enough to make me really rethink my health habits. I got on the wagon and was completely sober for months after that.

Sweden

1
Financial Ruin

Back in the United States, I would fall into the mainstream in the coming months. I believed the lie that I needed to search for a career immediately after college. Now that I had taken three months off to travel, it was time for "real life" to begin. I knew I needed somewhere new. I needed to find new friends and create a new life for myself if I was going to change my bad habits. I found a job in northeastern Oklahoma, working with cattle.

I had always wanted to move south, so I loaded up my old dodge diesel pickup and headed down to a little town north of Tulsa. I stayed away from addictive substances and slowed my nicotine use as well. I worked for three or four months. My job was seven days a week, it paid terribly, and my body was still not feeling right. So, I packed up yet again and headed back for Wyoming.

About five hours into the drive, the transmission in my truck blew up. It stranded me in Wichita, Kansas. I tried for days on end to find solutions. I did not have enough money to fix the broken truck. If I sold it, it would sell for pennies on the dollar, thousands less than what it was worth. Everything I owned was inside the truck, and towing it back to Wyoming was too expensive. I was defeated and broke.

A man I had met only once heard about my issue. He offered his truck and trailer to tow my pickup home if I could get a friend to drive it down to me. I found two girls who were willing to drive his rig. One was a friend I had known all my life, the other a great friend from college. They drove down and picked me and the truck up. I drove all of us back to Wyoming. I had another friend throw a new transmission in my truck for cheap, and my life was slightly back on track.

The man who offered me the use of his truck and trailer asked for a favor in return. I did a few weeks of ranch work to help him in a bind. During this trial period, he hired me full-time, and I shot up through the ranks in his company. He was a self-made man, not much older than me. The more time I spent around him and his family, the more I began to think he was who I wanted to be.

I knew a fair amount about ranching. However, I knew nothing about business or thinking outside the box. He taught me lessons that I would use for the rest of my life, lessons that would shape me. I eventually rose to the position of unit manager. I was controlling a couple thousand head of livestock spread across tens of thousands of acres. I was twenty-three and on top of the world. I was living my dream.

I had five men who worked under me, all Peruvian sheepherders. I got to speak Spanish all day, which helped my Spanish further improve. I even invested in my own livestock with my newfound confidence and knowledge. I went all in. I sold everything I had. I bought a six-hundred-dollar junk car to drive around, and that was all I owned. My little brother had some school issues, so he came to work with me for the summer. We lived together.

My brother and I racked up almost one hundred hours working every week. Finally, after almost a year of this, my boss, my mentor, and the man who had saved me in Kansas showed up at my house on one fateful Sunday in July. That morning, he delivered the news. Because of financial issues and my own shortcomings, I was fired. He gave me one week to get everything I owned out of the house I lived in. All my livestock had to be gone.

I had nowhere to go, so my brother and I packed everything we could into my crappy car. We hauled it to an abandoned barn to store it until we had somewhere to go. We lived in my car, two men, two dogs, and most of our stuff. We cooked on a little propane stove and used the trunk as our table. When possible, we stayed on some of my friend's floors.

When my one week was up, I needed to get my livestock out of there. I rented a truck, loaded them into a trailer, and hauled them to the sale barn. At the sale barn, I received less than I had bought them for. I lost thousands. I spiraled into the deepest depression of my life. I hammered the whisky bottle, smoked cigarettes all day, and felt bad for myself. My brother went back to live with my parents. I found a little work with one of my old bosses.

One day, I received a phone call from my best friend. He had found a truck for sale with a camper in the back. It was cheap, but it would be a place to live. I had just enough money left to buy it. We picked it up, and I could not believe how good of a deal we had gotten.

I began living in the little camper. I called it home. To this day, I still live in that camper. I love telling people it is an awfully slow vehicle but an incredibly fast house. I decided I needed to make massive changes in my life. If not, I would go completely off my rocker due to how I treated my body.

I had met a girl named Lilly from Sweden a few months earlier. We met in a ranch management course. I invited her to see the ranch I was managing at the time, and she, in turn, invited me to her family place if I was ever in Europe. My attempt to lead a stereotypical life was crumbling, so I returned to what made me happiest: travel. I made plans to go to Sweden and work for her family. Before the heavy winter set in on my Rocky Mountain home, I flew to Sweden for my next adventure.

2
Arriving in Sweden

Europe had always been low on my list of places to go, not because it did not sound interesting to me, but because it was so modern. So many tourists visit there that it seemed like it wouldn't be a challenge compared to the other continents. I always told myself I could go to Europe when I was old. I must travel to where the wild things are, sleep in the dirt, and tear up my body while I am young. I was not totally wrong. Europe would prove to be the tamest and easiest continent I would travel to, but easiest did not mean easy.

I flew to Copenhagen, Denmark, then took a train across the small channel that separates Denmark from Sweden. When I arrived at the train station, I met my friend Lilly. Her brother Karl, who I had not met yet, was with her. Lilly was all dressed up and wearing a nice dress. It was cold and rainy. I thought it was an interesting choice of clothes, but she brought it up first. She asked me if I could guess why she was dressed like that.

I guessed, "Because you wanted to dress up to meet me?" She laughed, "No, it's because I went out with my friends last night. I have not been to bed yet." It was about midday. When she said that, I was a bit concerned. I had been sober for an entire month since my life derailed in Wyoming. I thought I had walked into another party scene where I would party all the time for the next three months.

This was not the case, in fact it was quite the opposite. Lilly told me she was going to the gym the following day before the sun came up, before our full day of working cattle at the farm. She invited me to go to the gym with her. I said I would try it, but I had not worked out in years. She took a class every morning; the routine changed daily, and I happened to show up on a cardio day. Thirty minutes later, I was outside the gym's front door, puking in the grass.

I contemplated never going back to that place again. However, I knew it was offering me the life change I needed, so I stuck it out. While in Sweden, I went to the gym at least five days a week, a transformation that has stuck with me since I left. I continue to work out regularly because of the decision I made there.

The first day on the farm was interesting. Large cattle operations are not common in Europe. Cattle operations that do things "the cowboy way" are entirely unheard of outside Lilly's family. Lilly's father had grown the business into what it was. He had spent time in Australia, Canada, and the United States learning about cattle and ranching. He turned around and brought those techniques back to Sweden. He had grown his cattle business into one of the largest in northern Europe.

We rode horses almost every single day to move cattle; nothing motorized was used to move cows. Using horses for cattle work was uncommon in this part of Europe. The family could not buy horses for this work, so they had to train their own. Lilly had spent months working with horses and trainers around the world that dealt with livestock and equines. Through her experience, she had learned to train horses well. Lilly made some of the few rare, well-trained cattle working horses in Sweden.

The farm was on a large military base. They did not own the land; they simply had a long-term contract with the military for forage control. As far as the military was concerned, the cattle looked nice and kept the grass short. Well-maintained fields were important so they could use the land for training; that was all that mattered to them. Lilly took me first to look at their cattle. She wanted to get my initial impressions of some cattle they were about to sell for processing. We walked into a large green pasture full of white and tan cattle. When they spotted us, they came galloping over and surrounded us. They were like big dogs.

From a distance, I could not tell, but when the cattle got closer, it shocked me how large they were. They were much taller than typical American cattle bound for the meat processor. Tall cattle were the norm in Sweden. Many people thought the taller, the better. Laws, regulations, markets, and government intervention were different here than in most places. These factors made Sweden one of the most unique places to raise cattle in the world. The tall cattle we inspected were in a small paddock with sides made of portable electric fences. The family moved the electric fences daily to give the cattle fresh new feed. This was a progressive grazing technique I had been experimenting with in the United States before I left.

This type of grazing is like feeding a child. For example, if you set an entire birthday cake in front of a child, they will eat until their stomach is full and then destroy the rest of the cake on the plate. However, if you cut off a piece and give it to the child, the kid will eat until they're full. Then, there is little left to make a mess out of, and the rest of the cake is saved for later. This grazing technique used the same idea: only giving the animals access to a percentage of the whole pasture ensured the grass was consumed efficiently and evenly.

3
Cattle Systems

Since most cattle operations in Europe were small, they had limited equipment. A producer could not run into town and buy the supplies needed to run a large cattle operation like they could in the USA or Australia. Incredible ingenuity had come into play when the family had designed their facilities. Many factors had to be taken into consideration in the design. Rain and mud were always a problem in southern Sweden. They had to bring in many truckloads of gravel so the water would drain and keep the pens from turning into a muddy mess.

When I was younger, I got caught on horseback in a rainstorm with an old cowboy. We shivered and trotted back to the horse trailer, getting soaked to the bone. I made the mistake of complaining. He turned to me and said, "Cowboys don't get rained out. They get rained on." It was a statement that stuck with me through the toughest weather for years to come. However, in

Sweden, they did what they could not to be rained on; they built a large roof over the entire cattle working facility.

The facility had to be built by hand, piece by piece. Anything made from wood had to be resistant to moisture so it would not rot. Anything made from metal had to be resistant to rusting. Since a pre-made system could not be purchased, they hand-set every post. They cut every board to length and screwed it in place. The gates and metal sections of the fence were fabricated right there on the farm in their shop.

The gates had to be built so a person could easily open them without getting off their horse. This was easier said than done. Gate latches had to function perfectly and be easy to open and reach. As a result, ropes and pulleys were strung like spider webs through the roof. The ropes would allow a man on a horse to open and close almost any gate in the corrals with a pull. One person on a horse and one on the ground running the hydraulic chute could easily process hundreds of cattle daily.

Cattle in arid environments scrape their hard, dry hooves against rocks and gravel. This wears down their hooves, which helps keep them the right length. Much like a human's fingernails, their feet are always growing. In soft, wet environments like Sweden, the cattle have few natural ways to wear their feet down. As a result, cattle often came in with foot problems because of the moisture, mud, and overgrown hooves. This is also a common problem in dairy cattle all around the world.

The family had built a special chute to hold the cattle while they had their feet trimmed. This was done using chisels, rasps, and specialty knives. Recent technology, like the electric grinder, has also made things much easier. We would tie up one foot at a time and work on shaping and cutting away the excess, exactly like how a farrier trims a horse's hoof.

Sweden has an extensive cattle database. All producers are required to enter specific information into this database. We often put the cattle through the working facilities solely for data collection. As the cattle came into the chute, a scanner read the animal's ear tag. The scanner was hooked to a laptop beside the chute. Instantly, the animal's records and information would pop on the laptop. They equipped the chute with a scale that, within seconds, produced a weight for the animal.

They logged the information on the laptop and sent it to the government. All newborn calves had to be tagged and entered into the government database within ten days. If a cow died, it had to be recorded within that timeframe as well. Each animal had to have a tag in each ear, clearly stating the owner's name and contact info. If cattle escaped their enclosure, the owner could face steep fines. If the cattle did not have the owner's contact info, they could be euthanized.

The government was incredibly involved in every aspect of raising the cattle. They had all the possible information on the cattle they could ever want. Still, the government would send their own personal veterinarians to do surprise checkups. At any given time, they could show up and demand to test a person's cattle. Veterinarians checked on various animal welfare standards. The rancher had to stop what they were doing and bring the cattle in to be tested. They would collect urine to test for hormone levels, because all hormones were illegal in cattle in Sweden. They also tested that the weights reported were accurate.

It was illegal to collect semen from bulls in Sweden. This meant artificially inseminating cattle was not an option. This also meant that no samples could be collected for semen analysis, so it was impossible to tell if a bull was fertile. The only other method available was to separate each bull and give him a small group of females. The producer would only find out nine months later when he checked to see if any calves had been born.

Lilly's operation was far too big to be giving each bull its own group of females. So instead, they just sent out many bulls into a big pasture, hoping that there were enough fertile bulls to account for those who may not be fertile. It was highly likely that at least some bulls who had been on the ranch for years, eating and costing money, had never produced any offspring.

While I was working in Sweden, the family purchased a new piece of property. This property had no cattle working facilities, so they had to outfit it themselves. They did not build expensive and time-consuming permanent corrals. Instead, they opted for portable corrals. Portable corrals are extremely useful. They can be almost as strong as permanent corrals. Unfortunately, heavy-duty portable cattle panels were not a common item for sale in Sweden. Nobody in Sweden produced these corral systems. The ranch had to order them from a fabricator in Canada.

The corral arrived in a shipping container. Before they could assemble it, it had to be galvanized to prevent rust. I was told it took more than a year from when they ordered the corral to when it arrived. In the USA, a corral like this would likely arrive within a week or two of placing the order. We assembled the entire system using large tractors and four or five men. It took almost a week to assemble. We ran our first bunch of cattle into the corrals. It needed some tweaks to be perfect, but it was fully functional.

4
Bull Buying

I arrived on this Swedish ranch at a pivotal time. As the father grew older and the kids became old enough to manage things independently, he transitioned power over to them. Handing agricultural operations down between the generations is difficult. It has proven to be one of the biggest challenges to family ranches all over the world.

Recently, corporations have capitalized on this obstacle that family ranchers face. Usually, there are multiple children involved. Often, one or more children want their share of the business in cash. Siblings attempt to buy out their other siblings. Unfortunately, the business usually is not profitable enough to afford such a financial burden. I hear many people in the United States say this situation could very well be the death of the American rancher in the near future.

One responsibility already handed to the next generation at this operation was selecting new bulls to purchase. This is an enormous responsibility; the consequences of selecting the wrong bull can last years. If a bull with the wrong genetics is turned out with the cows, a producer will have generations of offspring carrying poor genetics.

Producers get paid by the pound. As a result, ranchers often choose bulls that produce offspring that will be as big and heavy as possible by the time they are sent to be turned into food products. Historically, this meant choosing the bulls that produced the biggest calves at birth. However, this method has started to change in the past decade.

Recently, producers have discovered that calves that are bigger at birth are often much less feed-efficient than calves that are smaller at birth but grow rapidly. Now, the desirable bulls produce calves that are small at birth, then grow quickly, so that they are heavy by the time they are processed. This has an added advantage because small calves make the birthing process much easier on the mother.

The likelihood of a bull producing calves like this can be determined by EPDs, or Expected Progeny Differences. EPDs are essential statistics. They are based on genetic material taken from the blood and entered into a database to compare to all previously-tested cattle. When shopping for bulls, the seller almost always provides the results of these genetic tests to the buyer.

I went with Lilly and Karl to look at new bulls available for purchase. We looked at many different breeds. Each breed offered its own genetic pros and cons. We planned to buy a few bulls of each breed and mix them all together. It would have been foolish for this ranch to put all their eggs in one basket and gamble on using just one type of bull.

Using one breed of bull is risky, mainly because the genetic pool in a country like Sweden is small. For example, if a large ranch in Sweden only raised Angus cattle for ten years, the ranch would quickly have every single Angus bloodline from Sweden in their herd. As a result, inbreeding would become a problem without strict genetic management.

Importing semen from other countries was an option, but it was expensive. Since hormones were illegal, it was hard to artificially inseminate the cattle with foreign semen. Without using hormones to synchronize the cows, multiple people would have to sit and watch the cattle for a month straight, waiting to identify the right moment for each individual cow in the herd.

Lilly, Karl, and I drove all over southern Sweden looking at bulls. It often surprised the people selling the bulls to see an American show up. We played games with the buyers, telling them I was a big-time American investor. I am unsure how many believed it when they inspected my shit-covered, tattered clothes.

We looked at Angus, Charolais, Simmental, and mixed-breed bulls. Since most cattle operations in Sweden were small, most sellers only had a handful of bulls to pick from. These bulls were often raised in barns like dairy cattle. They rarely had access to large, wide-open spaces like in other countries. The genetic component was an enormous factor, but genetics meant nothing if the bull had poor conformation due to growing up in a barn.

Bulls with poor conformation develop problems quickly. Poor conformation often leads to bad joints or a generally injury-prone animal. We had to avoid these types of bulls at all costs. When bulls are put out with the cows, they must cover a great distance to find all the cows they need to breed. The distance they cover will easily cripple a bull with joint or foot problems. Mounting and breeding a cow is also quite strenuous on the bull's body. As most people know, bulls also like to fight each other. Bulls are often injured or even killed when fighting other bulls in a pasture. Bulls with physical weaknesses will just not work on a large ranch.

The bulls we looked at often had foot problems due to living in barns and small pastures. We had to determine if this problem was caused by poor conformation or if the bull just needed its feet trimmed. If the bull needed its feet trimmed, we could handle that. When that bull was released into a big pasture, it would likely never have that problem again.

Sweden had strict traffic laws dealing with trailers and tractors. Unless you had a special license and permits, you could not pull a trailer. Even with the right license, the only trailer a person could pull with a regular truck was a small trailer designed to fit two horses. If we were going to haul anything substantial, we had to haul it with an eighteen-wheeler or big tractor.

When buying bulls, we brought the little horse trailer with us. We could fit one or two bulls in there and take them home. However, if we bought more than two bulls, we had to pick them up with a tractor pulling a large livestock trailer. This sometimes meant driving all day in a slow tractor to pick up cattle. So, when we looked at bulls far from the ranch, we tried to buy only two.

5
War Cattle

It was easy to forget that this cattle ranch was, first and foremost, a military base. Much of the ranch was just beautiful, rolling, grassy hills and forest. Public access roads were strung across the pastures. Cars drove by, often slowing down to inspect what these cowboys were doing in the middle of Sweden. I worked with one man named Åke. Åke took the cowboy lifestyle seriously. He was one of the few men in Sweden who had a large diesel pickup and a gooseneck trailer. On the side of the truck and trailer, it said, "COWBOY FOR RENT."

Åke's truck and trailer were standard for an American cowboy to own. However, in Sweden, he must have had them custom-built or imported. He invited me to his house to look at his American cowboy trophies. I thought it would be a few token items. But, when I saw his collection, I was surprised. It would have made any American cowpuncher proud to see. This man had spent incredible amounts of time and money accumulating gear. Saddles, ropes, bits, and spurs from the vaqueros of the western United States were packed into his house.

Åke and I often moved cows together. I would look across the herd and see a man in a flat brim, pencil-rolled cowboy hat. He rode a Wade tree saddle with custom "buckin rolls" bolted to the side. Long tapadero stirrups hung low to the ground, and a long reata was tied to his saddle. Big jingling spurs hung next to his horse's belly; he was something out of a Californio cowboy movie. He rode with a slump in his back, as many cowboys do after years of riding; his back had been made worse by a horse flipping on top of him some years before.

Like Lilly, Åke had to train his own horses because it was impossible to buy a working cow horse in Sweden. His kelpie dog followed along behind his horse. The dog was a fantastic working dog; it did anything he wanted it to. It was the only useful dog I saw during all my time in Europe. It was easy to forget we were in the middle of Sweden; I had grown up around men, horses, and dogs like this most of my life. However, Åke was the one and only Swedish cowboy.

Lilly, Åke, Karl, and I gathered a few hundred head of cattle: it was time to take this group to the working facilities. We had to give the cows and calves a checkup and enter some information into the government database. This would be a long drive; it would take most of the day to move them that far. We pounded along at a long trot in a straight line like the Calvary. The frost was heavy on the grass. It gave each step a nice crunch.

None of us had brought enough clothes for this brisk part of the morning. We knew it would warm up soon, and then we would have too many layers of clothing to take off. Until it warmed up, we just had to keep moving. When the large, yellow cows spotted our horses, they sounded the alarm. The cows went running in every direction, trying to find their calves. The cows knew horses meant they were going somewhere new.

We accumulated a large bunch of cows and calves and pointed them toward the dirt road. Once they were on the road, they would be able to walk faster. There would be less chance of them disappearing into the thickly-wooded areas. We made good progress that morning. The cattle had done this many times. They knew the drill.

Massive white calves trotted alongside their mothers, many of whom weighed well over a ton. A bull named Turbo also lived with this bunch of cows. He was one of the biggest bulls I had ever seen. He was a Simmental bull, brown with white spots. His hair curled up on his neck, making him resemble a buffalo. He was Lilly's favorite bull. She had shown him in cattle competitions when he was younger.

We came to the toughest part of that day's cattle drive. We had to get the cattle through a short stretch of a dense forest without losing them. Blackberry vines tangled the forest floor, making it hard to chase cattle if they ran away. We pointed the cows toward the trees. I rode in the back, in the drag position. They reserved this spot for the new guy who didn't know where anything was. The front positions had to steer the cattle; they knew the ranch well.

The horses in the front looked alarmed, as did the cattle. They could smell something in the woods ahead of us. We began applying more pressure from the back. The cattle pushed ahead into the thick forest; Lilly and Åke went with them. Karl and I pushed from the back. Things turned south just as we had gotten the last few cows into the forest. The cattle and horses spooked. I could not see what they feared. The trees blocked the sunlight. It was hard to see in the dim light.

I thought I could make out a figure crouched down behind a tree, a person. I could only see a silhouette. Then, it moved slightly. It was a person. They had a rifle. What the hell were they doing hiding? Then, almost on command, men began popping out of the forest everywhere. Slowly, quietly, they stood up. They all wore camo. They carried military-grade assault rifles.

The cattle and horses really freaked out when the men rose. I was a little unsure myself what was about to happen. I followed the lead of the others and we kept the cattle creeping forward. The cattle just walked right past the armed men, then we passed them on our horses. The men began saying hello in Swedish as I passed each one. They were soldiers. I had forgotten we were on a military base. We had just accidentally driven a herd of cattle through a war exercise.

We drew closer to the corrals. We were back out in the open now; nobody was going to sneak up on us. I now noticed there were many military vehicles out in the distance, green six-wheel-drive troop carriers. We passed a bunker where the troops were practicing throwing hand grenades. I thought I heard a low, powerful grumble in the distance. As I listened, I noticed it was getting closer.

A few calves had wandered away from the group. I went after them, forgetting about the rumble. I turned them back to the herd with my old palomino horse and trotted back to my spot. The rumble grew louder as I returned; it was just on the other side of the hill from us. Then, BANG. A tank came launching over the top of the hill. It tipped forward, smashing into the earth. The ground shook. My horse spooked. Black smoke billowed out of the tank as it headed away from us at full speed. I had never thought about what it would be like to drive cattle right through a war. Well, now I had an idea.

6

Bull Roping

Sweden had many rules, restrictions, and regulations around ranching. The family I worked for had fought a multi-year lawsuit against the country of Sweden over the right to keep their cows outside. They won the lawsuit. Before they won, it was illegal to have a cow outside in the wintertime in Sweden. Every cow needed adequate barn space to be kept inside all winter. Natural protection, like trees and rocks, did not apply. Man-made structures like windbreaks and open-sided barns did not apply either.

Where I am from, temperatures regularly reach forty below in the winter. Our cattle do fine outside in this weather when provided man-made or natural structures like the ones I have mentioned above. Southern Sweden does not get anywhere near this cold. Still, cattle were required to be confined to barns in the winter. The family fought this lawsuit because this was not feasible to do with over a thousand cows.

Since Sweden is a more socialist country, the tax is incredibly high compared to everywhere else I have ever been. The taxes make it awfully expensive for an employer to have many employees. Everything is done with the absolute minimum labor required, the opposite of most agricultural operations I have worked for. To keep the labor to a minimum, lots of machinery is used instead. Due to the laws around hauling, additional tractors are also required since pickup trucks may not pull heavy loads.

This ranch had more brand-new tractors than I had ever seen. Roughly ten large, technologically advanced John Deere tractors were used often. It was cheaper to buy automated equipment than to hire additional people to work on the ranch. It was fascinating to operate these machines; I felt like I was operating a giant robot.

We built miles of permanent and semi-permanent fences while I was there. It took half the time it would normally take me in the United States. The machines almost entirely removed the hard, physical labor aspect of the job. We had an excavator that put all the posts in the ground and vehicles to string up all the wire. Working in third-world countries taught me to work hard and value good workers. Working in Sweden taught me to work smart and value efficient machines. I try to implement a mixture of the two ideologies in my day-to-day entrepreneurial ventures.

Swedish cattle producers try to be as efficient as possible with the animals as well. Therefore, almost any type of cattle handling they did could usually be done with just two people. Cattle are often not branded in Sweden. If they are, they must have a heavy dose of painkillers. The added cost and time required when using painkillers makes the process inefficient. Swedes hate inefficiency, hence the legal requirement for very descriptive ear tags in every single animal.

The act of roping an animal of any size is also illegal. Even roping out of eyesight of a road is risky because private farmland is open to the public. A farmer cannot stop the public from walking or riding their horses across their property if the public is not destroying anything. This posed a problem when we had a sick or injured cow in a large, wide-open pasture.

If you cannot rope the animal to give it the help it requires, then more drastic measures must be taken. Often, this means stressing the animal out by cutting it out of the herd and making it walk to the nearest handling facility. Cows rarely leave the herd willingly. This means a lot of running, chasing, and stress for the already compromised animal, which in my opinion is far more stressful than roping. The only other alternative is to bring in all the animals together just to take care of the one or two animals who need help.

Bringing in a large herd requires multiple people and can sometimes take all day. Swedes hate inefficiency, and this method is as inefficient as it gets. Livestock producers also hate to see their animals suffer from sickness or injury. Putting the animal through the added stress of a day-long walk to the nearest facility makes the situation worse for everyone and everything involved.

While out checking cattle, we came across Lilly's favorite bull, Turbo. Turbo had lost weight. He was working hard, breeding his big bunch of cows and fighting other bulls for dominance. Weight loss is common this time of year for bulls. Ranchers often try to keep the breeding season short so the bulls do not overwork themselves. When we got closer, we realized Turbo had injured his shoulder, likely while taking care of his reproductive duties.

Turbo would be fine, but he needed rest. Chasing women is hard on a man. To give him rest, we needed to get him out of this huge pasture and take him back to the house where he could be alone. Lilly and I could see without a doubt that Turbo was in no condition to walk all the way back to the house, so we needed to load him into a trailer.

Loading cattle from a large pasture into a trailer can be one of the wildest experiences on any ranch. Sometimes, it can be done using cattle working facilities. In this situation, we had none nearby. If a person has some excellent working dogs and the cattle are not wild, the dogs can load them. We had no dogs of this caliber to assist us. That meant the only way to load him was to throw a rope around him and try to get him in the trailer.

Lilly and I returned to headquarters to get supplies. Lilly jumped in a tractor that pulled a large, custom-made cattle trailer. I grabbed a couple of ropes and headed back out ahead of her to find Turbo again. Turbo weighed one and a half tons. I knew I did not stand a chance of roping him off a single horse. So, I took the four-wheeler and tried to think of creative ways to capture him while I drove over.

When I approached Turbo, he began slowly walking away from me. His limp was visible, and he could not move fast. . I ran along behind him, swinging my rope. When I was close enough to him, I gave it a throw and, to my surprise, it landed cleanly around his massive neck. It occurred to me then that I had made no plan. I did not know what the next step was. I had just roped a bull on foot that could easily drag a car.

Lilly was slowly coming down the road with the tractor, but she would not arrive in time to assist me. I did the only thing I could think of: I sat down and wrapped the rope around my waist. It gently pulled tight, and Turbo continued to walk. I dragged behind him at a snail's pace. I believe Turbo did not even know I was behind him, slowly getting hauled across the pasture.

Lilly finally reached us with the trailer. I started thinking of ways I could quickly tie the rope to the trailer and stop Turbo. She pulled up and began rolling with laughter as she saw me getting towed around by Turbo. She turned the tractor off and opened the trailer doors. Turbo stopped walking and turned to face us.

I think only then did he even realize I was there. The massive, buffalo-looking bull then began walking straight toward us.

I stood up, unwrapped the rope from my waist, and began looking for a safe place to hide should he attack. Lilly said in a sweet, calm voice, "Come on Turbo, let's go home." Turbo walked up to the open trailer door and stepped inside without persuasion. Lilly slammed the door behind him. She jumped in the tractor, fired it up, and headed home.

7
When in Rome

They gave me a few weeks' vacation from the ranch in Sweden. I began traveling around Europe. I spent a week in Rome. Rome was supposed to be this fantasy land, like one giant museum. I had forgotten that nearly three million people live in that city. It's dirty, loud, crowded, and full of pollution and crime. The first morning I woke early and walked down to the Colosseum all alone. I took my camera. I wanted to capture some sunrise photos and some photos in the soft morning light, maybe beat the crowds as well.

It was a truly magical morning. My mind swirled with thoughts of gladiators, chariots, death, honor, and prestige. Lost somewhere in a fairytale in the back of my mind, I wandered past the Roman forum. I imagined all the ancient people that had once walked the same paths and explored the same buildings. I walked around the Colosseum, and a tall man from Senegal tricked and robbed me. It ruined my whole special morning.

I returned to the safety of the hostel I had booked. It was a happening place full of young backpackers, and it had a restaurant and a bar. I was given a top bunk, which wasn't my first choice, but it was all they had. A young, blonde woman from Chile came into the room. She had the bunk beneath mine. We spoke half in English and half in Spanish. It was her last day in Rome and my first day. I asked what I should try to see, what she had enjoyed.

She handed me a tattered map of Rome that they handed out at the front desk. It was covered with her notes. I realized she had mostly been to art museums. I looked closer at her and saw paint on her sweater; she was an artist herself.

She began explaining to me, throwing Spanish words into her broken English sentences, "It's not about what you're looking at. The brush, the pen, the charcoal, the clay, none of that matters in the big picture. Each piece makes you feel slightly different. So you must learn to interpret what the piece makes you feel and nothing more." She pointed out that many people in our generation had lost the ability to interpret feelings, or feel anything at all.

Most young people spend their days staring blankly at a screen, hanging their emotional coats on hooks made from likes and shares. Going to an art gallery can be as much of a journey into yourself as it is into new art mediums or history. This beautiful, blue-eyed Chilean woman had dropped a stick of dynamite into my head and walked away. I never saw her again; I do not even remember her name.

The newfound knowledge I had gained about how to look at art transferred into many other parts of my life: how I looked at nature, how I looked at cities, and how I looked at people, films, and relationships. I observed how I was reacting to each one of these things. Stepping on a horse. Gathering a pasture full of cattle, which I had done hundreds of times over the years. I found an incredible beauty that I had not seen before.

This new mindset made it quite easy to eliminate toxic people from my life that were doing nothing but sucking energy from me. It also made me stop looking at every beautiful thing through my phone camera; I just experienced it instead. Putting down the camera presented me with a problem I still battle with to this day. I love photography, but I hesitate whenever I pick up my camera and start snapping pictures of cowboys from faraway lands. Pictures will be special, but nothing can be as special as actually being in the moment.

While staying at the hostel, I met another person who would change my life in a different way. It was late, a weeknight, and nobody cared. We were all travelers with no time restrictions. We partied late into the night in the underground nightclub attached to the bar. Lights flashed everywhere along the ancient red brick walls. Electronic music bumped our brains into submission

I danced with girls who spoke languages I had never heard. I spoke with people from almost every continent. Many of them asked the same question. "What are you doing here?" It always caught their attention when I did not give them the typical answer. People always leaned in to listen closer when I told them I was trying to work as a cowboy on every continent.

A fiery Italian girl threw her drink in a man's face and slapped him hard right in front of me. The man had done nothing to her, he was just in the wrong place at the wrong time. Someone had grabbed her, and she had mistakenly thought it was him. He was from Mexico. As I laughed at his misfortune, he laughed too, and we struck up a conversation. He introduced himself as Carlos. When I told him what I was doing there, he told me his father owned a large farm in the state of Chihuahua, Mexico. He invited me to visit anytime I wanted. He would speak with his family, and whenever I wanted to go, I could just tell him, and he would take care of the rest.

I returned to Sweden, packed up my things, and said goodbye to my Swedish friends. I made plans to fly to Dubai for a week and check out some sights; I wanted to experience some Middle Eastern food and culture. Then, I boarded a flight bound for South Africa.

South Africa

1
O.R. Tambo

My journey in Africa started at O.R. Tambo Airport in Johannesburg, South Africa. It was the most interesting layover of my entire life. I had been in contact with an organization called the Future Farmers Foundation, and I had traveled here to help mentor and work with their students. Some of their students knew almost nothing about agriculture and were just getting started. Other students were running massive, multimillion-dollar farms. I had been told that a man would meet me at the airport. He was to help expedite my process through security and customs. I had trouble finding that man. I kept going at my own pace through the humdrum of airport lines.

When I finally had gone through all the security and had my passport stamped, they thrust me out into the real world. Porters swarmed me, asking to carry my bags, and taxi drivers yelled at me to follow them. Hands touched my body, hands touched all my bags, and I bolted out of the mob. I found the counter where I would need to recheck my bags and get my next boarding pass.

As I walked up to get in line, a young man wearing nice clothes and a tie approached me. He asked me where I was flying. When I responded "Pietermaritzburg," he looked worried and said, "Follow me, quickly. Your flight will leave soon, and there's no time to wait in line." I spun around and followed him. We quickly printed my boarding pass at a self-check-in kiosk.

He directed me to the security line, saying, "One last thing, sir, they ask us to check that you have at least 1000-rand cash. We do this to make sure you can stay in the country. Can you show me that much cash?" I had exactly that much in my pocket. Normally, I would be suspicious of this type of question. However, the week before I had visited Dubai, and they had asked me the same question, so I thought nothing of it. I pulled the money out of my pocket and watched as he counted it.

A man wearing a blue jumpsuit and carrying an AR-15 appeared and tackled the man with my money. The two men had a heated exchange, and the security guard asked if I knew the man. I replied, "No, he just said he worked for the airline. He was helping me go through quickly." Then, with a disgusted look, he looked back down at the man on the ground and began yelling at him in Zulu. I asked the guard if I could have my money back. Now even angrier, he continued yelling at the man on the ground.

I could not understand anything they were saying. This was my first time hearing Zulu; I did not know a single word. In fact, I did not even know the name of the language until a few days later. I was told by the guard to hurry, or I would miss my flight. The guard and the thief exchanged glances and a few words that let me know the guard was in on it now. The guard would let the thief go after I left. I was a little bit shocked. I was also worried I would miss my flight.

I had been picked up in a whirlwind as soon as I landed, and I was being thrown to the ground now. My phone was not connected to Wi-Fi, so it could not adjust the time since my last location. I honestly did not even know what time it was where I was standing. This made it harder to know if I was late for my flight. I rushed through security only to realize I had not even checked my bags, which were way too big to go with me in the cabin.

When I got to my gate, I saw I had almost two and half hours until my flight boarded. I was furious and broken-hearted. I had just lost all my cash, and I had given it up easily because I had been worried I would miss my flight. To be fair to myself, the flight was worth much more than the cash in my pocket. Now I had to go back through security to properly check my bag. In my initial confusion, I had even walked through security with a knife in my bag, without being stopped.

When I arrived at the front desk to check my bag, I told the ladies working there what had happened, and they sent me to talk to the police. The police said I had no time to file a report before my flight left. I would have to file the report when I landed. This was yet another lie. I gave up on getting that money back; it was long gone. I walked outside to get some fresh air before going through security again.

While outside the airport, I watched a young couple fight. The boyfriend had been in charge of watching their bags while the girlfriend was in the bathroom. Someone had walked by and taken all their belongings while he played on his phone. I felt like I was lucky in comparison. Then, a man with a sorrowful face approached me, holding a newspaper. He told me he had come from quite far away to interview for a job, and they had canceled at the last second. He asked if I had any spare change so he could get a taxi home. I told him I had just had my money stolen, otherwise, I would have been happy to help him. He genuinely looked sad for me and walked away.

Just a few minutes later, a different man walked up and told me the same story, and I realized it was all just a trick. I decided I would have to be much more guarded and less trusting of people until I learned otherwise. It seemed like everyone was out to get me; I stuck out like a sore thumb. I ran back into the airport, through security, and onto my flight. Hopefully, my contact would be waiting for me when I landed at the next airport. All would be right then. I would be in excellent hands.

2

Dairy Farming

My first opportunity with the Future Farmers Foundation was to work with a man about my age named Lelo. He was one of the first students to go through the program. Using the Foundation, he had found his way to Australia. In Australia, he worked alongside a dairy manager, learning all the skills needed to run a successful dairy. He worked there for over a year, perfecting his English and honing his management skills before returning to Africa.

Lelo had returned to put his new knowledge and skills to work, enriching his community by being a young black man in a management position. The owners of the farm were of English descent. When I spoke with them, they told me Lelo was the first black manager they had ever had, and they were incredibly pleased.

Labor in Africa is quite cheap. A thousand-cow dairy in the United States may have four or five full-time workers. A thousand-cow dairy in Africa would easily have fifteen or more people working simultaneously. Because of the large amount of labor, a manager's job was more difficult. It required not only farm knowledge but also exceptional skills in managing time and people. I only had a few days to spend on this farm and learn what I could. It was my first look into the real South Africa, and thankfully I had an expert guide to decipher what I saw.

Lelo had grown up in a poor, rural area. These were often the poorest parts of South Africa. Since he had started at the bottom and come nearly to the top, he could explain to me everything in the middle. Lelo explained that he could speak four languages fluently, and could easily communicate in a few more, which was common. Eleven languages are spoken in South Africa. To communicate with staff, everyone had to speak more than one language. I was amazed. Many of the farmworkers I met while in South Africa could hardly read, yet they still spoke three or four languages. They learned just from listening; it was a necessity.

Lelo would wake up early, as all dairy farmers do, and collect his staff to ensure they were doing their tasks. They milked the cows on a large rotating wheel, which made it easy for the staff and the cows to get the job done quickly. I had worked around dairies in my high school days, but this was my first return to the industry since then. I was quickly reminded that dairy farmers are some of the hardest-working people in agriculture. It seems as if dairy farmers do not sleep.

The assistant manager of the dairy was also a product of the Future Farmer Foundation. She was a young woman born and raised in a big city. Her family was a part of the large Indian population that had migrated to South Africa. Hers was a success story like nobody had ever heard.

As a manager on a farm, one of the biggest tasks is transporting workers so that they can be at the right place when work needs to be done. This means spending a big part of the day driving around with five or more men in the truck bed. All the time in the truck gave me a chance to speak with Lelo about his view of Africa and his past. Lelo previously had a semi-successful career as a rapper. He played me some of the music he had made. Lelo still had the occasional weekend show where he would rap for old time's sake.

In the United States, a success story is often referred to as a "come-up," a pull yourself up by the bootstraps, hard-knocks journey to the top. It was becoming apparent to me that the young men and women I worked with all had incredible come-up stories. They had come further than almost any American I had ever met.

I did not always get the opportunity to hear people's stories. But, without fail, the ones I did get to hear always inspired me. I don't think I really appreciated what I had been given in life until I heard about other people's pasts. I had started out quite low on the totem pole by American standards. However, I was now surrounded by people who would have given up everything for a chance to start where I had. That was humbling.

This realization made agriculture seem so much more powerful to me. Agriculture was a way out for these people; it was a path to a better life. It was very rare to see that. In most other places, working in agriculture was a choice people made purely for the love of the industry. In many countries, working in agriculture meant passing up on better-paying careers. This motivated me even further to try and mentor them as best as I could. I had chosen this field out of pure love, but they were using it to get themselves up and give themselves a better life.

Though I only spent a few days on the diary, it was truly one of the most eye-opening experiences I ever had, partially because it was my first immersion into South African culture. It was like an ice bath first thing in the morning: uncomfortable and shocking at first, but something that really sharpens the senses after the adjustment period is over. Living in South Africa was like waking up and wiping the sleep out of my eyes after twenty-four years, seeing the world as it really was for the first time in my life.

3
Meeting Batman

Before my trip, my contact at the Foundation told me that she had mentored a young man my age who she knew I would love working with. She could tell that our personalities would mesh. He was an incredibly hard-working young man, and one of the biggest success stories to come out of the Future Farmer's program. He was managing multiple large farms for a powerful landowner.

The young man had started his own livestock business as a side hustle and was doing well. I knew this was a guy I would like to learn from and hopefully I would be able to teach him something in return. I love working with other young entrepreneurs, people who have the hustle and drive to do whatever it may take to get something done. In my mind, this is one of the most empowering qualities a person can have.

I planned to visit this young businessman, whom they called Batman. I would spend about a month with him on the farm, both teaching and learning where I could. Time is a little bit funny in Africa. Nothing runs on time, not even close. So, when planning things, it was foolish to expect them to happen on schedule. This is exactly what happened when I was scheduled to meet up with Batman. Luckily, when I arrived at our meeting point and found he wasn't there, a neighbor offered to let me stay on his farm until Batman was available.

The neighbor had worked with the Foundation before, and had taken on many interns. I spent a day wandering around his large farm, which had crops as well as dairy and beef cattle. The neighbor's father offered to show me around and tell me some stories. It was fascinating to hear how the farming community was changing. It was even more interesting to hear how the country of South Africa had changed in his lifetime.

Finally, I got word that Batman was available. I threw my bag in the back of the truck, and we went to find him. The old man told me he had never met Batman before, but he had heard tremendous things from the neighbors about him. We met on a stretch of gravel farm road. We all got out of our trucks and shook hands. The old man pointed at me and said, "This guy is a little like a stray dog. He carries all of his things on his back. He will jump into anyone's car who has a place for him to sleep." It was a funny reality that had not occurred to me before.

Batman let out a loud, contagious laugh, which got us all laughing. I jumped in Batman's truck, and we headed off to his farmhouse. He was a single man, so when we walked into his three-bedroom house, it was almost empty. Like most bachelors, he did not even own enough stuff to fill this large house.

I threw down my bags, and we left right away. There was a lot of work to do, and work would not stop just because I was visiting. Batman had about forty employees under his command. Most of the employees required constant supervision from Batman, or no work would get done.

I was blown away by the size of the farms Batman managed. Tens of thousands of acres of pasture were his canvas. He filled them with cattle and sheep like an artist, moving them strategically, utilizing the available grass without affecting the ecosystem in the ground below. He handled all the livestock health, ordered all the feed, and mixed all the feed rations. He moved the livestock to new pastures and did everything else a good livestock manager must do.

Batman was the same age as me and had a high school education. Less than a decade ago, he had known almost nothing about farming. He was now managing a multimillion-dollar livestock business all by himself. I understood what all the hype was about, and why everyone had such good things to say about him. He was killing it.

On a large operation like the one Batman was running, there are constant problems that need to be addressed. For example, redwater is a disease carried by ticks that can kill animals within a day or two. If an outbreak happens and someone does not handle it immediately, an entire herd can die. There is also the constant threat of livestock theft.

The farm was close to Lesotho, a tiny, landlocked country inside South Africa. It was common for people to come out of Lesotho at night and steal livestock. The thieves mostly stole sheep. They would drive them back high into the mountains, where they would never be seen again. Because of this, every single animal had to be counted every day. The farm also had mounted guards for protection who lived with the animals. Of course, it was Africa, so there was also the threat of predators eating the livestock. Managing predators required tracking and killing them at times. These problems were not specific to Batman's farms, just the problems that any South African farmer dealt with almost daily.

I was also there during the peak of the South African farm murders. Many farmers, usually white farmers, were being targeted and robbed or killed. People living on remote farms had nobody around to help. The reasons behind these murders go much deeper than I can explain in this book. However, I was constantly asked by black and white farmers, "Why would you want to come here when all this terrible and dangerous stuff is happening?" It was a question I battled with myself.

Before arriving in South Africa, I honestly did not even know that the murders were happening. I met many farmers who told me their neighbors or family members had been killed in farm murders. The stories were truly horrifying: mobs arrived in the nighttime and brutally murdered elderly couples and kids.

Each night, we had a strict protocol for locking down the bars on all the windows of our house. The house also had a prison cell-style door that secured all the bedrooms. The courtyard walls had barbed wire running along the top. A private security company was on call 24 hours a day. With one phone call, they would show up with guns blazing. I suspect we could have easily been trapped inside if a fire had started; it was a fair trade to ensure criminals could not get to us.

4
Death and Danger

Danger was all around me the whole time I was in South Africa. It was something I was constantly aware of, but it rarely changed the course of my day. At this point in my journey, I had begun to realize how much my outlook on fear had changed. I noticed that there was a lesson to be learned every time I was afraid. Sometimes, the greatest lesson to take out of that fear is to witness it and learn to remain calm. My thoughts became clearer, yet my brain still was still able to rage inside with possibilities.

Before my world traveling began, I chased that feeling of being afraid. Looking back, I did not handle it well when I found it. I constantly sought out fear and discomfort. I made poor decisions that would have repercussions for years to come. I was reckless; I sailed my ship over many unknown waterfalls for no reason other than to see the other side. I had matured with my travels, but I had much left to learn, and South Africa was going to show me that.

I had not yet dealt with death. I had dealt with danger, but not enough to see death in person. Ready or not, it came screaming into my life in Africa. Life seemed to have a cheaper or more replaceable feel here. Maybe they did not sweep death under the rug and hide it like in wealthy countries. Many industrialized countries do this, so that the world feels more comfortable. I am thankful I got to experience the opposite.

It was explained to me that many Africans fear dead bodies and the power that they contain. Because of this fear, paramedics would not load corpses into ambulances. If locals saw a corpse being loaded into an ambulance, they would refuse to go in that ambulance; they would associate it with death. Refusing to go in an ambulance is a choice that can kill an injured person. So, when there was a dead body, nobody touched it, and they waited for the morgue to pick it up, which often took a long time. These bodies sat out in the open for hours. It was not uncommon to be driving somewhere in Africa and have to swerve to miss a dead body surrounded by grieving family members.

Not everyone lives a long life in Africa. This especially displayed itself in the thinking of the lower-income South Africans I met. The mindset was once explained to me by a group of young farmworkers. These workers had a morbid yet beautiful outlook on life that completely shattered my paradigms. The men were making fun of one of their coworkers for saving money. This coworker saved half of each of his very modest paychecks. He was the only worker on the farm who owned a car, yet these men still saw no point in why he saved his money and lived frugally off of bread and water.

The men told me, "If I die tomorrow, every bit of my money left over goes to my family. If my family gets my money, they will spend it all within one week, throwing a big party where everyone gets drunk. No matter how big my savings, they will spend it. Now, if I spend all my money, at least I know if I die tomorrow I will have died with a full belly and one last cold beer."

The idea of constantly spending all your money sounded ridiculous. I was taught from the beginning to scrimp and save. However, I continued to think about their point of view. You can't truly disagree with someone until you can see and understand the reasoning behind their argument.

I still do not agree with spending your money like there is no tomorrow, but it taught me a valuable lesson. I always strive to live in the moment, to be present in what I do every day. This outlook on money made me realize that, in their own way, many poor Africans were living a fuller life than the richest people in developed countries. This is because they do not just handle their money as if each day could be their last: they handle every aspect of life that way.

I believe this makes them incredibly happy people. They likely go to the grave more satisfied than most people I know. However, I do not believe this is the perfect way to live. I am now in my mid-twenties. Many of my friends are making forty-year retirement plans. I often wonder if they could not use a bit more Africa in their lives; I wonder if they forget to live today. They are living forty years from now in a future that they, for some reason, think is promised to them. Many Africans know with a full heart and clear conscience that they are not promised any time.

I have a business degree from an American land grant university. I have owned a few businesses of my own. I probably read more business books than the average person. Yet, somehow, one of the most profound financial realizations of my life came from this conversation with these farm workers, while shoveling six inches of smelly muck out of the bottom of an old stock tank the size of a swimming pool, in well above ninety-degree heat.

5
Jackal Hunting

In my time with Batman, I learned more than I ever expected. He taught me far more than just techniques used in ranching: he taught me about taking pride in your work. Both Batman and his staff took extreme pride in their work. Many of his employees were from the country of Lesotho. Some of them had entered South Africa legally, and some had not.

Lesotho guys were my favorite to work with because they were exceptional horsemen. Most South Africans feared horses and would rather walk great distances than ride. The men from Lesotho came from a herding culture high in the Drakensberg mountains. They grew up on horses, and they could ride and train them well. One young man from Lesotho named Kuda quickly became a good friend. He began teaching me about his country and his home.

The daily stock counts took a large part of the stockmen's day. Sometimes the counts were quick and easy. Other times they had to cover a vast distance, gathering every single animal to get a perfect count. They became quite good at counting because if they came up one short on their count, they had to gather them all up and do it again. They did it until they got the right number, or else they had to prove someone had stolen some animals.

The men that rode horses loved their horses. Even though the men had few resources, they were willing to give most of that to their horses. They would make things out of grass, sticks, and trash. For example, I saw one man using a dog food sack stuffed with wool as a saddle pad to cushion between the saddle and the horse's back. It was crude, but it worked well. The horse never got sore. They rode in cheap fiberglass saddles, which were extremely uncomfortable for me, but the men got used to them. The saddle resembled a pack-saddle tree made from cheap, thin fiberglass.

They gave all the employees company clothing. The truth is, most of them had little clothing of their own to wear while working. All the men wore tall rubber boots to keep their feet dry. They were often ill-fitting and would tear up the men's feet. The men on horses would ride in these rubber boots, and the men who did not ride would walk miles and miles in these boots.

The men who would not ride horses probably walked more than a dozen miles a day. They walked through rain or shine, hot or cold, always searching for the livestock in the big mountain pastures. The sheep were the hardest to watch. It is hard to steal a cow, but sheep are easier because they are small. During my time there, we only had problems with sheep theft.

The men at the border patrol had become friends with Batman. The border patrol would call him anytime an animal with his farm's brand went across. Horses were returned this way, as well as sheep and cattle. Sometimes they would steal fence posts and other equipment. Batman had to paint everything that could be stolen so he could prove it belonged to his farms when they found it. Batman would dump supplemental minerals for the animals, using big truck tires like bowls for the minerals. Once, a man tried to ride over the border into Lesotho on his horse with a tire wrapped around him. They caught him and returned the tire to Batman.

Another thing that became a serious problem was jackals, which are African coyotes, or dingos. These jackals rarely bother a sheep. Jackals never bother cattle, but when livestock have young babies, they come around and raise hell. Jackals would kill a few lambs a night if we did not do something about them, so Batman and I took a few measures to stop them.

First, Batman hired a crew to put Jackal netting around the outside of the Lambing corral. Jackal netting is a fine woven wire Jackals cannot fit over, under, or through. They also cannot chew a hole in it. This made a big difference. Batman also had men take turns on the night watch for predators. The last resort was jackal hunting. Batman hired a crew of professional hunters with an enormous pack of dogs to track and kill jackals. One of Batman's co-managers was a big hunter, so we took his rifle and did some nighttime hunting ourselves.

Jackal hunting with forty dogs was a hell of a time. One man ran with the dogs on his horse, and the other men trotted behind the dogs while tracking the Jackal. About half of these dogs were hound dogs with good noses to track the scent. One-quarter of them were greyhound-type dogs for chasing. The last quarter were Jack Russell-type dogs that would go down into the tunnels. In the tunnels, they would fight the Jackals underground. All the dogs were pretty much wild. They hardly listened to a command. If the owners did not tie them all up, they would find something to kill.

One day a hunter called and asked us to bring a shovel. He had found a jackal in a burrow. When we got there, I stayed on top of the truck. Batman said it was best I stay up there. The Jackal hunters all echoed the same statement. They said the dogs had only ever been around black men; there was a good chance they may bite a white man. Whether or not it was true, I was certain this pack of savage dogs had not seen a white man. I was not going to find out about that theory. The men dug up the Jackal, and the tiny dogs bailed into the hole to fight with it.

The hunter grabbed the back leg of one terrier named Smoothie. Smoothie had only one eye. When he pulled on Smoothie's back leg, he came up with the Jackal clamped in his jaws. The other dogs attacked the Jackal as soon as they saw it. With one swift swing of a cane, the old man killed the Jackal, and he fought to get it back from the pack of dogs.

It was wild and impressive to see how the canines worked together as a team to hunt. Each breed of dog had a specific purpose. Even with very little training, they instinctively knew what to do. It was raw and simple, and it made me think about how people had hunted before guns, how man and dog would have started to become best friends thousands of years ago.

6
Welcome to Africa, Mate

In Africa, I saw more dead people and violence than I had ever seen before. Guns were extremely hard to get legally. It often took years of paperwork just to acquire one gun legally. Getting a gun illegally was far easier to do. The men who carried guns usually carried them illegally. These men were often men to be avoided. If someone was willing to openly carry an illegal firearm, they were also probably willing to do something with it. Bus stations and bars were often the scenes of many violent crimes, sometimes involving guns or knives, other times just sticks or fists. Even some businesses participated: the taxi companies in South Africa had fierce territory wars that often became deadly.

One night I was with some friends, and we went to buy beer. Upon entering the bar where we would buy the beer, I received a pat-down from a private security officer. Peanut shells and drunk old men lay scattered across the establishment's floor. Loud African music played over booming speakers. When I entered, everyone stopped and stared. If there had been another white man in this establishment before, I imagined it had been quite some time.

I walked with my friend to the counter to buy what we needed. He looked around the place and got an uneasy feeling. He turned to me and said, "Maybe it's best if you wait in the truck with the driver." My friend was a strong Zulu man. Places like this did not scare him. He understood I was the outlier; this could turn dangerous.

I returned to the car outside. I sat with our driver in silence. We were on a dark, narrow street, directly opposite a bus stop filled with drunks and dangerous men. Just then, a drunk man approached the driver's side door. He began speaking harsh words in Zulu and reaching his hands in his shorts. I sat in the backseat, unseen by this man. I saw his hands in his pants, and I leaned forward to tell the driver to watch this man's hands.

Before I could get the words out of my mouth, the driver said to the man, "Come closer, I have something to show you." The man leaned his head in the window to see what it was. The driver pulled a pistol from the waistband of his pants. In one fluent motion, he cracked this man over the head with the gun. The drunk man stumbled off into the darkness, holding his bleeding head. The driver said nothing.

Shortly after, my friend returned with his arms full of beer. I told him what the driver had just done. He shrieked with laughter and turned to face me in the back seat. With a grin, he said, "Welcome to Africa, mate! Now let us go get some KFC." The fast-food fried chicken was a favorite of many in South Africa. It was a rare treat to get into town and eat some chicken.

We sped away into the dark night; Zulu rap music pounded through the truck speakers. Soon after, we sat in the KFC lobby, deciding what we should get. I looked over to see our driver with a 1911 pistol still hanging out of his waistband. He spotted some good-looking young women. He smoothly approached them to get their numbers as we ordered food.

This was just a normal Saturday night in South Africa. It had gone well because nobody had gotten hurt. Unfortunately, this was not always the case. While living with Batman, one neighbor had seen thieves stealing sheep in the middle of the night. He pulled his hunting rifle from his bedroom and shot one thief in the back, killing him. He was desperate to get rid of the body; he took a tractor and attempted to grind the body into the ground with a farming implement. It did not work.

The farmer turned himself in. The police are so slow in South Africa that they rarely get things done. It took them four days to go to this man's farm to look for the body. When they finally did, the body had been taken. They had no evidence, so they could not charge him with anything, and they released him.

When I was finished working with Batman, I headed north to work on another farm. I bummed a ride with an old cattle trader. We planned to stop along the way because his sister owned a coffee shop about halfway there. After about ten minutes of visiting his sister and finishing our coffees, we hit the road. Further down the road, we came to a roadblock. We sat in a traffic jam for 15 minutes. We were one of the first cars to hit the roadblock, and the line of cars grew long behind us.

I could see a large truck stalled in the roadway a few hundred meters ahead of us. It looked as if it had lost some cargo. Finally, a police officer approached us and began telling all the cars to turn around, find another way. The road would not open for some time. When the cattle trader asked what had happened. The cop replied, "These idiots tried to rob a bank truck in broad daylight. Just before you stopped, there was a huge gunfight." What I thought was spilled cargo was bodies of the dead bank robbers scattered around the highway.

That night on the news, I saw the report. The police had gunned down twelve or thirteen men just moments before we arrived. The robbers never even breached the armored truck. Thankfully, nobody was injured besides the robbers who died. They had fired thousands of rounds in the gunfight. The casings lay scattered around the highway as a reminder. I wondered what would have happened to us if we had not stopped for that ten-minute coffee break. We likely would have been right in the middle of that gunfight. Guns, theft, and death. The side of South Africa nobody wants to see.

7
Farming in the Northern Desert

A fter driving all day, we finally arrived in a dusty little town in the north of South Africa. It was a small farming town that sat near the Orange River. The Orange River was the lifeline for most agriculture in the northern part of the country. Crop farms lined each side of the river, and huge irrigation pumps brought the water up out of the river. They spread the water across the land using modern pivot irrigation technology.

I had been told to meet up with a man named Bertie. Bertie was another star in the Future Farmers Foundation. He had helped many young people gain farming experience; he had many interns that worked on his farm beside his normal hired staff.

In this part of the country, they almost exclusively spoke Afrikaans, a language similar to Dutch. Many stone bunkers stood out in the arid hills surrounding the small town. These bunkers were left behind after the Boer war, a fierce war fought between the British and Dutch settlers in the early 1900s. Ties to this war are still strong. British-Afrikaans tension is always present.

Bertie was a young man, not much older than me. He had spent his younger years playing in a rock band named Zinkplaat. Then, he settled down on his family's farm with his wife and child and took over from his father. His father had an incredible farming career. He was once named the South African farmer of the year.

When Bertie returned to the farm to take over from his father, he decided he wanted to make some big changes. He would slowly transition the farm from conventional farming to organic. This transition is not something that can be done overnight; it requires careful planning. The transition was less than halfway completed when I worked for Bertie.

Bertie owned a house in town that had multiple bedrooms. I would stay there with two others who had also recently started working for Bertie. One of my roommates was a young man named Simon. He had studied at Stellenbosch University near Cape Town, the most prestigious university in South Africa. Simon had come to learn about organic farming. He planned to have his own farm one day.

My other roommate was a girl named Motlao. Motlao had plans to manage a farm of her own someday as well. She had been taking University courses, and now she was getting the hands-on experience she needed. Motlao could speak ten different languages. South Africa has eleven official languages, and the only one Motlao could not speak was Afrikaans. However, that would change over the coming months as she worked with Afrikaans-speaking employees.

Together, the three of us would commute from the house in town out to the farm. It was a half-hour drive each way. The road was dusty and rough. We would ride in the back of a truck along with some of the other workers who lived in town. Kudu and other wild game were prevalent in this part of the country, so the driver had to always be aware. Our driver was an assistant manager who lived in town. He had worked for Bertie's family for years.

It could be chilly driving to work in the early morning. We often huddled tight behind the truck's cab to stay out of the cool wind. The wind was our best friend by the time we were driving home in the afternoon: temperatures were above 100 degrees by then. Dripping with sweat and covered in dust, we would load back into the back of the truck and stand as the truck sped down the bouncy road, letting the wind dry our clothing and cool our faces.

The farm had success transitioning to organic corn. They also planted organic pumpkins, peanuts, and other cover crops. I was not involved with the crops very often, so I did not get to see much of how this was done. The farm also owned Dorper sheep, pigs, some Boer goats, and my specialty, cattle. The cattle operation was a purebred Bonsmara operation. Bonsmara is one of the best cattle breeds in South Africa.

The family had developed a reputation for having some of the best cattle in South Africa. The Bonsmara cattle are deep red and handle the heat exceptionally well. They sold many bulls to farms all over South Africa. The bulls thrived in the hot, red sand. We fed them big sacks of whole kernel corn that the farm had grown. They stayed fat and healthy off this corn and the shrubs that grew in this hot, dry place.

The farm also raised yearling cattle to be processed for meat. I spent much of my time helping them work on the intensive grazing system they had developed to feed these cattle. On one of their circle-shaped pivots, they had planted corn and a mixture of other cover crops. To the average person, it probably looked like a cornfield that had not been weeded. This cover crop was perfect cattle feed.

We took electrical wire and built semi-permanent areas to contain the cattle. Every few days, we moved the cattle to a new area of feed. We would build small, sliver-shaped areas for them to graze, like slices of pizza. The idea was the same one used in Sweden: give them a small amount of feed, and they clean it up with no waste. This grazing practice has a much more positive impact on the biology of the soil and plants.

The cattle would often escape the electric fence as we figured out the best ways to contain them. The young workers and I would spend hours running in the heat of the day looking for a few renegade cattle that had escaped. The cover crop was twice as tall as I was and it was dense. We would find the cattle by searching in a grid pattern. Often, we only heard them at first because we could not see them. The boys yelled directions in Afrikaans. I quickly picked up pieces of the language.

We used long, machete-like tools to clear our way through the crops in search of cattle. We also had to slash out clearings for the electric fence. If the crops touched the wire, it would ground out, and the fence would be nothing more than a thin string. This was often the reason the cattle escaped when they did. To build a new area, we had to slash plants twice as high as a man all the way down to knee height. This job gave me big, painful blisters on both hands. Swinging the machete all day in the heat took its toll on a man.

8
Runaway Horse

Time and time again, I'm reminded why I stopped telling people I know how to ride a horse. In other countries, that usually meant one thing: "Can you take this wild, angry, spoiled, half-psycho horse that nobody has touched in three years and get a day's work done on it?" This was the case in northern South Africa.

I had ridden some horses on Batman's farm, moving sheep and cattle. Those were horses that had been ridden down. They were used and trained nearly every day. In the wide-open deserts of northern South Africa, horses need to travel long distances quickly. These horses differed greatly from the ones I had ridden in the south. They were much taller, with long legs and necks, and they only knew one command: "GO!"

The farm did not own horse trailers, and the farm was so large that it was too far to ride everywhere. So instead, the farm kept pairs of horses strategically placed in different places, so they could be caught and used on each part of the farm. I am unsure if this is a common practice in northern South Africa; maybe it was specific to the place I worked.

The problem with having these horse pairs all over the farm was that they rarely got used. The horses would probably spend months out in a big paddock before being used for one day, then released again. The horses quickly forgot all their training, and they learned terrible habits like under-ridden horses do. They had all but reverted to wild horses by the time you needed to use them again.

I did not know this about the horses until after the fact, so I jumped at the opportunity when asked if I wanted to help with some cattle work on horseback. We caught the two horses by driving around the massive, flat paddock, chasing them with a pickup and honking. Finally, they got a bit tired and stood in a corner so we could catch them. This was my first sign that things were about to get a little western.

We saddled the two feral beasts with the cheap fiberglass saddles, so uncomfortable to ride in but cheap enough that the workers could afford them. I saddled one horse as he walked in circles because it was impossible to make him stand. He was hyperactive, and we had not even started working yet.

Together, a man named Michael and I swung up on these half-wild horses and took off. I had brought my nice camera and had it slung over my shoulder, which was a mistake. The camera began bouncing as soon as my horse moved at any speed. I worried about it breaking, but I needed both hands to control this horse. There was nothing I could do about it now. Both horses gave us little bucks; they reared up on their back legs, and we both had a brief runaway within the first few minutes.

After some time, the horses seemed to have the bucks out of their system. They were still oozing with energy as they began sweating heavily on that cool morning. They would only power walk or prance, even prancing sideways sometimes. Someone had clearly bred these horses to run. They had received little training on how to slow down, it seemed.

After five or ten minutes of trotting in the cheap plastic saddle, I could already feel my leg rubbing raw from the nylon stirrup straps. When I asked Michael where we were going, he told me, "You see that hill there? We are going there to meet my cousin, and then we will look for some escaped heifers." At first, I had fixed my eyes on the closest hill to us. It was less than a mile away, and I thought I could put up with trotting that long. However, it turned out he meant a much bigger hill in the distance: this hill was probably four or five miles away.

My legs rubbed badly, and I was riding in tennis shoes. Twice already, my stirrup had bounced over my ankle, every horseman's nightmare. You are stuck if the stirrup gets past the heel of your shoe and up to your ankle. If I fell off the horse, I would be dragged, badly hurt, and possibly killed. This is why cowboy boots and riding shoes have tall heels.

Finally, I told Michael I could not take it anymore. I felt terrible saying it. I hate being that guy, but I could no longer take the rubbing. I could feel the blood running down my leg. Michael said we did not have time to stop. He had to catch up with his cousin, so he told me I could stay behind and ride slower. Hopefully, no lions or hyenas would eat me.

I held my horse back until Michael had gotten far in front of me. I barely had the power to slow this horse down. The horse fought the bit the entire time. We were riding on some very rocky, uneven ground. Going fast would be dangerous here. Many of the barbwire fences had a single hot electrical fence wire attached to them. This wire kept the wild game, ornery heifers, and horses from tearing through the fence. It carried a powerful shock.

When Michael had almost made it out of sight, his horse whinnied to mine, and my horse became frantic. I could no longer slow my horse down. I could not even steer the horse anymore. I was now on top of a total runaway. I did my best just to stay on top of the horse and not get my feet stuck in the stirrups. I nearly wet myself when we came to a steep, dry river crossing. The crossing was lined with trees. Like most plants in very arid environments, these trees had large, sharp thorns to protect themselves from predators.

My horse leaped down into the river, dashing through razor-sharp branches. We ran across the riverbed and exploded through the nasty trees on the other side. I emerged out the other side, still on top of the horse, but not looking good. The trees had cut me badly. My shirt was nearly torn off my body. I was bleeding from a thousand tiny cuts on my arms, chest, and face. Also, my favorite hat was missing. A good friend had given me the hat when I had watched his band perform. Now the hat hung somewhere in the branches that my head had smashed through.

I did not realize it, but the river crossing was easy riding. Now, my horse was running flat out over rocky hills. All the tricks I knew to stop a runaway horse were not working. It was far too rocky to bail off, so I pointed the horse toward Michael and his cousin, who had arrived. I tried to steer my horse around a particularly rocky area. The horse did not respond. We were totally out of control now. The horse crashed into a tall termite mound, smashing my leg and the horse's left hip.

This impact with the mound sent the horse's rear end skidding around to the right like an out-of-control rally car. The horse did everything it could just to keep its footing. Finally, the horse's rear end hit the fence line. Before it had a chance to hit the barbed wire fence, it hit the electric line. I heard it give the horse one hell of a zap, and we were off again at a dead gallop. Finally, we caught up with the others. You can imagine the look on their faces. We rode up bloody, my clothes half-ripped, and both of us dripping with sweat.

They felt terrible for leaving me behind when I told them what had happened. They even went back and got my favorite hat for me. They said it was still hanging high in the tree, exactly where my head would have been passing through. Thankfully, "Tris Munsick and The Innocents" was still legible on the hat; it was unharmed. They had also seen the blood and little pieces of clothing that had been ripped off me.

It was funny when it was all over, but I had been terrified. Michael and his cousin pulled out some small pieces of newspaper. They dropped some smoking tobacco in the paper and rolled it up into a smoke. They always used newspapers because they couldn't afford normal rolling paper. They offered me one. I turned it down. I watched the blue and green smoke roll off the end of the cigarettes as it burned the printer ink. I vowed not to ride any of those horses again.

9
The Auction Barn

During my stay in the northern deserts, I met many interesting characters. Everyone in this region spoke Afrikaans. However, they often spoke English as a second language. One farm manager had spent years working in the United States on a farm in North Dakota. I believe he took a liking to me because I reminded him of his younger days in the USA.

The man invited me to attend an auction with him where they would sell sheep, goats, and cattle. These were not just any livestock: they were some of the most high-end animals in the region. Naturally, I jumped at the opportunity. I would remind him at least every other day to make plans so we could go.

It was a long drive to the auction; everything was far apart in this hot, dry north. I learned a lot on that drive, particularly about the Dutch, Afrikaans, and Boer. I had recently read "When the Lion Feeds" by Wilbur Smith. Some say it is a biased book, but it offers a look into what South Africa used to be like economically and culturally. I could see how it brought up painful scars for modern South African society.

When we arrived at the auction, we went to make a left turn into the parking lot. An older black man walked right in front of the truck. We had to slam on the brakes to avoid hitting him, which nearly caused another car to hit us. The man I was with lost his temper. I saw that steam was about to come out of his ears. He calmly told me, "Had that happened in the old days, during apartheid, most people would have gotten out and beaten that old man down. At this auction, there are probably some people who still would."

We grabbed a sausage and a coke, then grabbed good seats close to the sale ring. Hundreds of animals had shown up at this auction, and it was open to the public. Anyone could buy livestock here: I had seen a sign clearly stating that as we came in. As I looked around, I realized that meant it was open to white people of any kind. If you were black, you were handling the livestock, it seemed.

Many men approached me as we walked around looking at the animals before the sale. Many of them assumed I was an American investor or someone looking to import South African genetics into my herd. These men approached with flyers and business cards, shaking my hand and giving me big smiles. They did not do this to anyone else there. The men really thought I was there to spend big money. Little did they know I was scraping by, bumming rides around Africa just trying to stretch my dollars as far as they would go.

Many of the world's most productive sheep and goat breeds come from South African bloodlines. They had developed these breeds in the same region where this auction was being held. So, as I walked around, I saw old breeds I was awfully familiar with but also new breeds I had never heard of. Anyone passing by was happy to tell me all about these new breeds. I could not help but think that I was seeing the future; these breeds would someday be internationally known. We took our seats as the auctioneer began calling us close so he could start the bidding.

The auction began like any auction in the United States would. They rolled in the best animals right at the beginning to get the money flowing. The first big ram they brought up had won many awards. He was a masterpiece, one might say. A big, fat man in the bottom row bought him for over ten thousand US dollars. People clapped and cheered, and they moved to the next one.

This went on all afternoon as they rolled in animals one by one and sold them to the highest bidders. The auctioneer would occasionally throw out some words in English I understood. By the end, I was getting a handle on the Afrikaans number system. Most importantly, I realized that in English, when we would say twenty-three, they would say three and twenty in Afrikaans. I could not understand Afrikaans, but the more time I spent around people speaking it, the more I heard similarities. Eventually, with some practice, I could understand basic commands and sentences.

We did not buy any animals at this sale. To me, it was all about experiencing this culture, seeing the farmers who battled in this tough part of the world. We loaded up in the truck for the long ride home. We had a few stops along the way to say hello to my friend's family. First, we stopped at his family farm. He showed me his crops, which had come in nicely. He also showed me his pigs and the goats that he loved. All his animals were fat and healthy. They thrived in this desert environment.

Someone from a wetter place would likely look out into the bushveld and say, "Nothing can live there, unless they eat rocks and sand." Those were exactly my first thoughts. Yet, herds of hundreds of goats and sheep somehow managed to feed themselves out there. This is why tough South African breeds have spread worldwide to extreme climates.

As we were looking at the animals, an old Toyota truck pulled up to the nearest gate and hammered on the horn. A skinny, old black man in tattered work clothes walked over to the gate and let the car through. The car pulled up, and an old white man stepped out. He walked up to us and began grumbling in Afrikaans as he shook our hands.

My companion stopped the old man and introduced me. He said, "Dad, this is my friend. He came all the way from the United States to visit farms here. The man only speaks English. You can't speak in Afrikaans to him." The old man looked surprised for half a second before turning to look right at me. He said, "I don't speak fucking British," and then carried on in Afrikaans. It made me and my friend laugh. I was not surprised. Many of the older people still carried that much hatred from the Boer war, a war fought over one hundred twenty years ago.

10
Hunting and Fishing

I had made many friends while I was working around the farms. Many of them would become friends I would talk to regularly for years to come. One of the best friends I made was a young veterinarian who was just a few years older than me. His name was Roche. I heard from multiple people that he was one of the brightest young vets working in South Africa. He could use ultrasound to pregnancy test animals with blazing speed like I had never seen.

Roche was a tall man with a thin but powerful build. He wore boots made from kudu and elephant skin everywhere he went. No matter the weather, he wore shorts year-round. Afrikaans was his first language, and he spoke impeccable English and Xhosa. He reminded me of a character from a Wilbur Smith novel I had read about the Witwatersrand gold rush in 1886.

When Roche came to ultrasound some of our sheep, I spent the better part of the morning talking with him. My questions began as veterinary-related questions. They quickly switched gears into questions about the outdoors and hunting. I told him if he was ever in Wyoming, I could hook him up with both hunting and fishing trips if he wanted. I had hunted and fished most of my life; they were my favorite activities.

Roche never missed a beat while accurately determining not only if a sheep was pregnant, but with how many babies. He only needed about ten to fifteen seconds per animal, and he did it all while holding a conversation with me. He invited me to go hunting or fishing with him in Africa if he ever got the chance. He usually scanned around thirty thousand sheep per year, plus thousands of cattle. That was just his pregnancy testing work; he did much more normal vet work.

Despite his busy schedule, Roche and I planned to meet up and do some African hunting while I was there. It took us some time to make it happen, but it was well worth the wait. I caught a ride out with one of the delivery trucks that worked for Roche's vet practice. They dropped me at Roche's home clinic. From there, I jumped in with Roche, and we made a few quick stops on the way to his house. We scanned cattle and looked at some of his neighbor's sick old dogs.

His house was very remote. It reminded me a lot of my home in Wyoming. Huge, steep hills surrounded the farm, which was located in the Eastern Cape of South Africa. The hills were covered in a rough, woody plant, much like sagebrush. There was little farm ground and not a tree in sight. It was all livestock grazing. I was to stay with Roche for a week or so, and I would get in all the hunting or fishing I could handle.

Behind his house was a small stream that he said was full of trout. He set me up with a fly rod, fly box, dirt bike, and two-way radio and sent me off to do some fishing. He told me that the two-way was the only way to communicate this far out, and that other people could always listen in on the channel. It reminded me of the stories when old-timers would tell me about talking on a "party line," when an entire district would share the same phone line in the rural parts of the USA. Everyone on the line could hear everything.

I hopped on the Honda dirt bike and skittered up the stream. After I stuck a wooly bugger on my line, I began walking the banks looking for big deep holes with large fish. I found a few that day. I caught five or six fish and was quite pleased with myself. I fished like this for a few days. Then, Roche and I loaded up his truck with hunting gear and headed out into the Karoo region.

We set off with his old .300, binoculars, and enough meat to feed a few men for a couple of days. We drove like banshees through the night, trying to make it to the hunting place in the dark. That gave us the chance to have long conversations.

At that time, being a white landowner was a little unsettling in South Africa. There was continued talk of land reclamation without compensation. Roche had bought his own farm just recently. He had worked hard for it. The government was now considering coming in and taking the farm from him and kicking him off the land. The issue is far more complex than this. Nevertheless, it was a terrifying idea as a young business owner and native-born citizen. After taking all the entrepreneurial risk and building a successful business, there was a chance it could all be taken in a flash by the government.

I spoke to him about my failed and successful entrepreneurial ventures. I also spoke of the tiny part inside me that worried about keeping up with society, and how I had already given up a lot of opportunities to travel the world. He then said to me one of the most powerful things I can remember hearing. He said, "Sometimes I would rather own a little and see the world than own the world and see a little." As we buzzed along the dark road, Kudu ran in front of our headlights. I stared blankly, trying to process what he had just said.

We approached the hunting camp in the dark. It was near midnight now, and we could see that a group of men had surrounded a large fire; they were also hunting on the same land as us. They were old friends of Roche's. We stayed up late into the night drinking Black Label beers before crashing into big tents. We had to wake in only a few hours, when the sun rose, to go hunting.

When I woke the next morning, there was a very thick fog. The burning orange African sun tried to pierce the veil that surrounded us. Roche and I began to quietly trek our way up the hills, always on the lookout for wild game. This kept my heart rate elevated. The fog finally broke around nine in the morning. We saw that we were standing at the edge of an enormous cliff overlooking a wide-open valley. From there, we spotted multiple springbok, but we decided they were not worth pursuing.

We hiked up the ridge and scared a group of springboks ahead of us. At three hundred meters in front of me, a beautiful male presented himself. I laid down to steady my aim. Roche watched the animal through the binoculars. I shot. The bullet soared just barely over the animal's back, but it was all over. He ran far out of sight. I would never see him again.

11
Hunters or Hunted?

After missing my first shot at 300 yards, we walked around the rest of the day searching for another buck, only to find nothing. I was heartbroken. I thought I had missed my only shot at a springbok. Roche told me we would go out again the next morning and try to shoot one when the sun came up. That night we stayed in a massive old farmhouse.

The house was owned by a man with whom Roche had gone to vet school. He was a hard-nosed Afrikaans bushman and veterinarian. His family had owned the land for generations. His parents had been raised in this enormous house, and now it lay all but delinquent. They visited the old house just enough that it would not fall apart and be forgotten.

We sat on the porch between massive white Victorian pillars that held up the front of the house.

We lit a small fire, cooked wild game meat, and sipped on some beers. We told stories about hunting, women, and the wild things we had seen. The sounds of Africa surrounded us. The cackle of a jackal or the scream of some bird or monkey would pierce the darkness, just to remind us what lurked in the darkness beyond the safety of our fire, beyond our big white house.

The next morning, we went out before sunrise. We saw a buck right off the bat with barely enough light to take aim. We snuck up remarkably close, maybe 50 yards away from him. I had never ejected the shell from yesterday when I had missed at 300 yards. I tried to remove the used shell in the chilly morning light and replace it with a loaded one.

Even in the low light of morning, my movements were enough to alert the springbok of our presence. He ran off before I could even fire. A rookie mistake had cost me that one. I thought I was cursed, that this hunt was over. We continued, not wanting to waste the precious golden hours of the morning. We spotted a buck from a distance and began our stalk on foot.

We trotted down a rocky ridgeline, hunched over to keep our profile as small as possible to avoid detection. My arms burned from two days of carrying around the .300 Remington magnum rifle. My feet were tired from the previous day's romping around the desert canyons looking for game. When we began the chase for this buck, my old friend adrenaline kicked in and took away the discomfort.

We got to a big, flat rock sticking up above the coarse, brown grass. I laid on it with my rifle in the ready position. The buck was slowly but surely coming straight toward us, so we waited and let him keep coming. The sunrise was a burning red mass by this point in the morning; the bleeding skyline flooded the mesas and canyons of the Karoo desert as we waited.

Then suddenly, through the silence, came a low rumble. I thought maybe it was a big diesel truck in the distance, but we were much too far from any roads for that. Another rumble came from somewhere in the distance. Then, finally, they came together as one massive rumble. Some sort of primordial fear jumped in my throat as I lay on that cold, black rock. Roche looked at me out of the corner of his eye, never putting down his binoculars.

I whispered to Roche, "Is that......lions?" He clearly heard the concern in my voice. With a smile, he said, "Yes, just wait," and so I did. The roaring built into a grand crescendo carrying across the wilds of Africa. The pride of lions was obviously large and close. I gripped my gun a little tighter. Shivers raced up my spine, and goosebumps dotted my arms. I counted how many bullets we had left. Only three. I decided I wanted to keep every one of those for fighting off lions.

I told Roche I did not want to shoot the springbok. I figured it would be like ringing the dinner bell for the lions. Surely, they must know that when a gun is fired, there must be meat and guts to eat. Roche quietly chuckled at me. He seemed all too easy about the fact that there were a bunch of wild lions just over the hill from us. Then, he let me in on a little secret. They were in a game reserve, and just out of sight was a huge fence that held them back. They could not get to where we were.

My breath still quivered. I could not just shake off the sound of lions roaring in the distance. I tried to calm down. I realized that listening to those lions would maybe be one of the most special moments of my life. Watching the sunrise with my rifle, lying on a rock: it seemed almost like something out of a movie. I refocused myself, looking through the scope and watching the springbok. The lions continued to roar for the next ten minutes.

The springbok was finally in range at two hundred yards. Whenever he turned broadside, I was ready to take the shot. When he finally turned, I made a nearly perfect shot. He ran twenty meters before dropping dead. Roche and I ran up with excitement, took a few pictures, and cleaned the guts out of him. Roche said we still had a few hours of hunting left; we could try to find a bigger one. So, we stashed the carcass under a bush in the shade and continued.

We got lucky, and the next valley we walked into had a bigger springbok at two hundred and fifty yards. The sides of the valley were very steep and covered in loose gravel and thick brush. I had to lie in an extremely awkward position to steady the rifle. I pulled the trigger, and he dropped, never taking a single step. Two springboks down in a thirty-minute time frame: I was elated. As we ran down the steep hill to retrieve our trophy buck, I saw something running through the thick brush directly at us.

We dropped low to the ground and watched in silence. Suddenly, about twenty mountain reedbucks jumped out of the bushes. The gunshot had scared them, but they did not know where it had come from. Now, they rushed at us. Roche whispered, "If there's a buck in this group, drop him." We sat still. The animals ran so close to us that we could nearly touch them. Even so, they did not see us. Finally, the last one to emerge from the bush was a big male.

I jumped to the closest bush, steadied the rifle on the biggest branch, and hurried to find him through the scope. He was moving fast. At last, I found him. He stopped for a split second. I put the crosshairs on his heart and pulled the trigger. We were not sure if I had hit him. As we scoured down the hill, we found him lying dead close to where I had shot him. Three bucks in a row. I had never imagined my hunt going this well.

We loaded all three in the truck and drove home. When we returned to the house, we hung them up to cool. The next day, I skinned them and removed their heads. Then, I began breaking them down. Roche helped me with the butchering process. We gave half the meat to the families who worked for Roche. Many of them could not afford meat.

The meat we kept I made into Biltong, a type of jerky. I cut the meat into strips, seasoned it, and hung it outside to dry for two weeks. I worried about how sanitary the meat would be. I still ate nearly every bit of it or gave it to my friends. Nobody got sick. This is how they traditionally preserved meat in Africa.

I packed up my bags as I had done so many times before and found myself a ride to the airport. My wallet was empty, and my heart was full: South Africa had been everything I ever wanted and more. It was a soul-shaking place filled with the most amazing people. I knew I would miss it as soon as I was home.

Mexico

1
Chihuahua

After I returned from Africa, I worked in the US for almost eight months on a ranch in the high deserts of Southern Wyoming. Finally, I decided I would head to Mexico for the winter. It was getting cold. Flights were cheap, and frankly, I had to leave the US for a while to maintain my sanity. I called my friend Carlos, the Mexican man I had met in the Roman nightclub. He was happy to help. He lined up everything for me and arranged my pickup in the city of Chihuahua.

It was unclear what type of work I would do or where I would live, but I did not care. So long as I was outside the US, my itch would be scratched. Carlos's sister, Maria, picked me up from the airport. She was young and beautiful, with long black hair, and she spoke English well. We got some tacos, and she told me a bit about the family and where we were headed. She had a deep love for Mexico and its culture. She was an incredible linguist.

She spoke Spanish, English, French, German, and several other languages.

As we headed to the farm, I asked her the question that had been burning in my mind, the question so many people had asked me before I left the United States: "Is it safe here? What about the organizations running drugs to the border?" She just laughed at me, but it seemed like an uneasy laugh. She reassured me that the place I was going was very safe. However, one of the most powerful men in Mexico had a vacation home in the town. He was the king around there, and that was all I needed to know.

I went back to meet the family. They were wonderful, pushing delicious food at me like I was starving. The dad began explaining to me what type of work I would be doing. He had huge apple orchards. They were also about to harvest corn. He said he had owned heaps of cattle until about six months before I arrived. He had sold them all, and now he only farmed.

They had family and friends with cattle. He said he would help me get work with horses and cattle. When I could not find cattle work, I would stay with him and work. He took me for a ride in his brand-new, fancy truck, maybe the newest in the whole town. We rolled out of the dusty little town to look at his cornfields. We tested the moisture in the corn. He showed me where his land started and ended. I told him a little about me and where I had been.

My Spanish was barely conversational, so we kept our talks brief and simple. We drove back into town and stopped at the store. The lady handed us six Indio beers in a plastic bag full of ice. We opened one each right there, "Salud!" and drove off to one of the old man's best friends' houses. We had a big cook-up: steaks, tortillas, and chilis, all cooked on a cut-up old oil drum grill. Just me and a bunch of old Mexican farmers. I listened to them carefully, trying to understand each conversation.

I was part of the good old boys' club now. I would see this group of old men all over town for the rest of my time in Mexico. They were always eager to help me or just yell "EY GRINGO!" with a smile. American people did not go to this town. I was probably the first in a long time. The town was not about vacations and white, sandy beaches. It was dust, one-dollar burritos, and skinny cattle.

I loved my new little northern Mexico desert town. The place looked like something out of the movies. Every few days, I would get a stark reminder of the danger that was always present. Federales would roll by fully-loaded, wearing masks and ready for a shootout. There was always an ambulance at the pub right across the street from where I lived, hauling out people with knife wounds. Nobody asked questions. I learned quickly that one did not ask about those in power. I kept my head down, and nobody bothered me.

They gave me an old beat-up Ford pickup to drive around while I worked. I picked up some other workers on my way in to work each morning. People always stared at me when I drove down the red dirt roads lined by mud-brick homes with tin roofs. I stopped most mornings to buy a few burritos from a girl with a roadside stand. This was my lunch every day. Every day in Mexico, I would eat beans, tortillas, and some sort of meat. They were always mixed differently or served in a special way. The type of food never changed, but it was still delicious every time. Some mornings I would get eggs. Occasionally we would mix in some homegrown chilis, and on extra special days, we would get cheese.

I worked with a whole swarm of guys while we were farming. They were always coming and going. Labor was cheap, and times were hard, so labor was easy to find and easy to replace. Some guys would just leave after one paycheck. Almost every one of the guys I worked with had crossed into the US and worked illegally at one point in their lives. They all said the same thing, "I made a lot more money there, but I just spent it faster." This is a problem I have seen all over the world.

Often, people from low-income backgrounds in any country are never taught much about money or financial literacy. Financial illiteracy, unfortunately, keeps some people unstable their whole lives, regardless of income. Most of my coworkers had worked construction in Arizona, where illegal immigrants are the only people they can get to work outside in the summer heat. Each of them told me they had returned home to Mexico to see their families, and then Trump was elected in the US. When they tried to sneak back into the US, they found that Trump had sealed the border up much tighter.

2
Cutting Corn

As the nights became colder in Mexico, the corn turned brown and lost its moisture. We had to wait for the moisture to drop to a certain level before we could harvest. We filled our days with mechanic work on all the harvesting equipment. I also helped with tile work on a new house out on the farm. But, mostly, we were killing time waiting for harvest. One of my first jobs was cleaning out an abandoned lot full of old metal and big rocks. We flattened it into a pad to park machines.

Everything we moved brought funny-looking bugs and snakes up out of the ground. I always had to watch where I stepped. I spent almost two days walking around and picking up stones as the other men flattened the earth. It was backbreaking work. At the end of the second day, an old man came up to me with a smile. I was covered in dirt and sweat, my hands raw from handling rocks. He said, "Now you're working like a Mexican!" It made me laugh. I knew all too well the type of labor my coworkers had been subjected to when they illegally crossed into the US.

The apple harvest had just ended when I arrived. I was sad to have missed the chance to see that. However, many trees still bore fruit that had not been ready at harvest time. I could walk into the trees and pick a few small apples for a snack when I was hungry. We cut and ground and welded on machines until finally, the corn was ready. The combine went lumbering through the town streets and out into the fields.

In the town, the power lines were extremely low. The combine was too tall to get under them. The old man told me to get up on top and hold up the power lines while he drove under them. Dangerous work rarely fazes me, but that I flat out refused to do. When I was young, I had seen a man hit a power line with a piece of equipment in the US, and he got killed.

One worker named Juan jumped up on the combine to do the job. He seemed totally unphased. While he was up there, the old man threw him a roll of electrical tape. He told him to patch up the weak spots in the power line because the government surely would not do it. I looked away, fully expecting Juan to hit the ground dead. He did it with no issues. The machine passed easily under multiple power lines with Juan's help.

The old John Deere combine reached the field. The driver dropped the head into the corn. He flipped the switch, and dust billowed out of the back of the machine. I already knew I was prone to hay fever, but this was my first time around corn dust. I would spend the next month sneezing and fighting allergies I had not expected.

Every morning I showed up with another young guy to service the machine. We filled it with diesel, checked the fluids, and started it by using a wrench to cross the terminals on a battery while hitting a solenoid with a hammer. My experience is with livestock, not so much machines, but I knew a few things about equipment. All the chains were loose and rusty. We lubricated them by simply dripping diesel on them every morning. I greased the bearings with an old grease gun.

I had to take a corn stalk and dip it in a bucket full of thick grease on chilly mornings to load the grease gun. Then, I would pack the grease into the gun using my hands. It was a very messy job, but I got good at it. Everything we did was crude and slow and had me thinking, "We would never do this back home," but I also found the process inspiring. Whatever problem we encountered, we just had to figure it out. That usually meant long hours and hard labor. There was often no easy way to do it with the tools we had. It was a satisfying feeling to know that we, and we alone, were responsible for our problems.

Huge eighteen-wheelers would show up to be loaded with corn as we harvested. They were maintained the same way our machinery was. The trucks had bald tires, and most of the lights did not work. Many parts were held together with wire and tape. The trucks would never have been allowed to drive on the roads in the US, but somehow, they kept them working in Mexico just fine. All the corn that spilled while loading the trucks had to be picked up by hand. That was my job, with just a shovel and bucket.

When I had filled a few buckets, I had to climb up the side of the truck, then pull the buckets up with a hook and rope and dump them. The combine also missed a lot of corn. When I was not shoveling spilled corn or working on a machine, I had to take a big sack and walk each row of corn looking for missed cobs. Juan and I walked hundreds of acres hand-picking corn. It was miserable.

One day we had a big screw-up. The combine driver dumped a bunch of dirty corn into the wrong truck. The factory we delivered to would not accept dirty corn. That meant that four guys, including me, had to take buckets up into the belly of the truck and begin throwing corn over the sides into another truck.

It took nearly two whole days to unload one truck into another, a job that would have taken ten minutes if we had a grain auger.

As I quickly learned, I am terribly allergic to corn dust, so I had to have two bandanas tied around my face while doing this job. If I began sneezing or coughing, I had to climb out of the truck to get fresh air for five minutes before I could return to help the other guys. They were all shocked that I had come from the United States to work with them and do this type of labor. Many of the young men in this area were scrambling in the other direction to get away from this. Many times, I asked myself the same question, "Why the hell am I putting myself through this?"

3
Firecrackers

Mexico is built up to be a big, scary place for many Americans and people all over the world. One of my African friends commented that he would never go to Mexico, too dangerous. His statement was surprising to me. South Africa is certainly just as dangerous, if not more dangerous; this was part of the attraction for me. I have found in my travels that rumors about countries are usually blown out of proportion. Mexico, however, turned out to be almost exactly as the rumors said.

When watching the news, viewers do not learn about all the wonderful people and the amazing small farming towns that fill the interior of Mexico. People only hear about the criminal activities that happen in Mexico, which is a shame. I googled the name of the town I planned to live in before leaving the USA. The first thing that came up was a news article covering the facts of a raid where explosives had been used to enter a building. The police extracted someone they believed to be one of the top operators in an organization that operated outside the law.

Most locals assured me I was in safe hands, that if I minded my business, everything would be fine, and they were right. Nobody ever bothered me even though I was the only gringo in the little farming community. Still, they kept a close eye on me. I often heard people say, "Does he speak Spanish? Do we need to worry about this guy?" When that happened, I would immediately remove myself from the situation as quietly as possible.

I occasionally saw huge rolls of money being passed between people near me. I did not question how they earned that money. My Spanish had improved drastically by this time, so I could understand most conversations. I believe this was the key to keeping myself out of trouble. There are things I saw that I have chosen not to include in this book to protect myself and everyone I was associated with.

One night after work, my coworker and I jumped in his old, beat-up SUV. It still had Texas license plates on it, a few years expired. He told me it had been stolen from the USA by someone else. My coworker had bought it from them cheap. The police in Mexico had so many bigger problems that they did not bother with stuff like stolen trucks. My coworker told me that many of the cars in town were stolen. I am not sure how true all this was, but these were the type of questions that, for my safety, I did not dig any deeper into.

We picked up some of his friends, all guys my age, and bought a box of small Indio beers and headed up the mountain. When we got to the top, we all cracked open our little beer bottles. We looked over the lights of the tiny town. We spoke of the futures the other young men wanted to pursue. All of them dreamed of making it to the USA legally. They also dreamed of owning their own small businesses or farms in Mexico, but unfortunately, Mexico had become too full of corruption and bribes for them.

Then, faintly, we heard firecrackers being set off in the street in the town below. We continued scheming and planning out everybody's dreams. Ambitions are something I love to speak with others about. Then, more firecrackers popped off, but closer and louder this time. These were different kinds of firecrackers now, some making deeper booms than the others. Just then, my friend's phone rang. It was our boss.

The boss gave stern instructions to "get the gringo home, NOW." We did not know what was happening, but the urgency in his voice meant to hurry. We sped down the mountain. We hit the dusty dirt roads of the town. Windows down, we blasted T3R Elemento on the radio, swerving to avoid the three-legged dogs and roosters that scattered under the flickering orange street-lights.

When we reached my house, the grandmother let me in. Her hands were shaking so badly that she could hardly unlock the door for me to enter. She hurried me into the house. I had to lock the door behind her. Her trembling hands would not cooperate.

When I asked her what was wrong, she said, "Listen, don't you hear?" We paused and listened to the silence of the night. I heard no cars or music in the street, which was odd, only the occasional dog barking. Then the noise was there again but awfully close, the sound of firecrackers going off. "Guns," she whispered in a shaky voice, and all at once it occurred to me. It had not been firecrackers this whole time but a gunfight going on in the streets.

Grandma told me to stay below the window level of the old house to avoid catching a stray bullet. She told me not to turn on any lights until morning. I thought I had misunderstood her. My Spanish was still far from perfect. I texted my boss's daughter in English to confirm. She assured me it was real.

The daughter said I should be genuinely concerned, but that by morning people would have cleaned it all up. I was not about to ask questions about it, so instead, I laid on the cold tile floor underneath my window. I listened to the chaos unfolding outside, protected by iron bars over the window.

Adrenaline rushed through my veins while I laid there trying to breathe myself into a state of calmness. I could not help but feel that same feeling as when I was riding out on the roof of that bull truck in Australia. I was scared for my life. I was scared for others. But, the fear did not paralyze me as it once had. I controlled my breathing. I just soaked it in and felt the terror. It was no longer an uncomfortable feeling, just a feeling.

The ruckus outside tapered down after about 15 minutes, and I fell into a deep sleep. The next morning there was no smoke, no bodies, and no bandits, like nothing ever happened. A friend said to me, "That is just the way they like it, nobody knows anything, and you do not dare ask."

4
Idiot

My boss pulled me away from the corn harvesting for a while. I was suffering because of my allergies. He knew that I competed in and loved rodeos. He had me work helping to set up for local rodeos. I dealt with the bucking stock and prepped the arena days ahead of time. By doing this type of work, I met quite a few locals.

The locals were absolutely stunned when they learned where I was from and what I was doing in Mexico. I earned the respect of many of them after dragging around a heavy fire hose in the mud all day, watering the arena. When the day of the rodeo came, my boss bestowed me a great honor by letting me ride his big, beautiful paint horse in the local Cabalgata. A Cabalgata is a cowboy parade. It starts in the next town over, then everyone gets together and rides their ranch horses right up to the rodeo arena in celebration.

They pulled a live mariachi band on a trailer in front of the parade. Behind the band was another trailer loaded with a few thousand cans of Tecate beer, so that a person could trot to the front of the parade and order beer, then trot back to their place in line and get hammered without ever getting off their horse.

That morning I was nervous. The horse had not been ridden in some time. They just gave me the horse and told me good luck. Nobody else I knew would be riding with me in the parade. An old man hauled my horse to the beginning of the parade for me. He must have seen the look of worry on my face. He insisted I do two or three shots of tequila or Mezcal before I got out of his truck. It was nine in the morning.

My stomach burning with courage, I jumped up on my horse to join this moving party. I was riding one of the tallest horses in the parade and one of the few paint horses. I stuck out like a sore thumb. People started noticing me, yelling "hey! Gringo!" and giving me a beer or cigarette. I quickly realized they intended on getting me really drunk.

I was becoming friends with the whole small parade. I ran into a guy my age who I had met at a party previously. He was a local rancher. We rode most of the parade together, talking about ranching and Mexico. He invited me to work on his ranch, an opportunity I jumped at. We planned for me to help him work cows so I could get out of harvesting corn for a while.

As the parade made its way closer to the rodeo arena, we passed through ranches and tiny little towns. More horses and riders joined us at every place we passed. We now had hundreds of horses with us. One man told us he was so desperate to join in that he had to use his son as a trailer driver. He pointed out his son to us as the twelve-year-old boy rolled past, navigating the truck through traffic, pulling a horse trailer.

It was midday now, and the sun was hot. Many of the people in the parade were totally wasted. We pulled into the fairgrounds, where everyone would get a big meal prepared by all the wives. We sat between giant concrete cones painted white that stood over a hundred feet tall. These cones were very rudimentary silos that they used to keep the grain dry after harvest. I had also seen the occasional cock fight there before. This was the end of the parade.

They instructed me to gallop my horse back to the ranch, unsaddle him, and get a ride back to help with the rodeo. So, I did just that. Everything fell silent as I rode away from the ruckus of the parade. I could now only hear hoofbeats and the occasional squeak of the leather saddle. I imagined I was an outlaw.

I had seen so many cowboy movies and read books about bandits running to Mexico to live in dusty little no-name towns just like this. I remember being in that moment for what seemed like forever. I am so thankful I got to be there, galloping home and feeling the wind on my face. It is a feeling I hope to never forget. Somehow, it felt like I could feel the energy of all those bandits and outlaws, both the real ones and the ones from folklore.

After I put my horse up, I went back to the rodeo. I brought my camera to document what I could from the little hometown rodeo. By now, I had stopped drinking. I was sobering up but still had a belly full of courage. It convinced me to get into spots that I normally would not to get great photos. I took some of the best photos I have ever taken at that rodeo. I shot photos of both the rodeo and the people attending.

Everyone was friendly to me and did not seem to mind being in the photos. That is until I made a mistake that, to this day, still gives me goosebumps thinking of how close I came to disaster. The rodeo had ten or more gunmen placed around the event. In my drunken ignorance, I assumed they were police, there to assist the public if something went wrong. The men looked very official, clean-cut, and they all had bulletproof vests and machine guns.

I watched one of the armed men crack open a beer as he began talking to one of the rodeo participants. Most of them were smoking cigarettes. That sort of thing seemed too casual for American police officers, but in Mexico, that seemed normal. I decided it would make a great photo to get this mixture of cops and booze and cigarettes and horses and guns all in the same photo. I thought I should ask first though.

I walked straight up to one of the men, who was leaning on a brand-new Mercedes. I later learned he was the boss of all these other guys, which was why he was guarding the car. I asked the man nicely if I could take pictures of him and the other guys with him. Quickly, he grabbed my camera and demanded to know if I had pictures of them. I showed him that I did not, and explained that I had planned to ask first. He looked me up and down like a piece of meat and then told me to go away.

I learned later that night that those men were not police. They were bodyguards, Sicarios for one of the most powerful men in Mexico who was in the stands watching. The man I asked for permission to take a picture had probably killed many men before. Probably stupid gringos, just like me. The moment when he looked me up and down like a piece of meat and saw right through me,I know he was deciding if he should kidnap and maybe even kill me. I will never know what convinced him not to. I must have caught him on the right day.

5
Pig Roast

A few days after the rodeo celebrations had settled, I went to work on my new friend's ranch. He ran a few hundred cows with his father. They mostly ran the cows on leased land. This meant moving cows every few days to a new pasture. They used a grazing system, much like the grazing in Sweden. They gave the cows small areas to graze for a few days and then moved to the next area so that they did not waste food.

Moving to new pastures frequently meant building a lot of temporary electric fences. However, there was a stark difference in equipment compared to in Sweden. My first job was to help build a new perimeter fence. They gave me a wheelbarrow full of hand tools and some posts.

We dug each corner post by hand. The ground was sandy, with many rocks the size of my fist. After the sun had been up for only a few hours, it was already ridiculously hot and dusty. We spent all morning digging holes and setting fresh-cut cedar posts in the holes. The sandy soil did not hold the posts very well. Each post only had to hold up a thin electric line for a few days, so we left some of them a little wiggly.

Next, we strung out the electric wire. It was wound around a beat-up old piece of plastic with a hole running through the middle. We stuck a straight stick through the hole in the plastic and walked. The wire rolled out freely as we walked along. We pulled the wire tight at each wooden corner post and tied it off. Next, we grabbed some fiberglass posts. Fiberglass does not conduct electricity, so we could hang the electric wire on these posts without worrying about grounding it.

With age, fiberglass frays and breaks. These posts looked almost wooly from the amount of curled-up fiberglass strands sticking out of them. I had handled these old sorts of posts before, and I knew better than to let them touch my bare skin. When picking up a post, I wrapped it up in an old piece of cloth. We lightly tapped the little posts into the ground fifteen paces apart and strung up the line. The last step was to add the electricity.

As with almost everything electrical in Mexico, the fence charger was sketchy. The wires had been broken many times and tied back together. The solar panel was cracked and dirty. I was surprised that it worked at all. I have great respect for Mexican ingenuity. These men could make anything work when it broke. We hooked the charger up and it worked, letting out a pulse of electricity every few seconds. Usually, people have a small tester to touch the fence that can read if electricity is passing through. Not in Mexico. We walked along the fence, touching it lightly every so often. Each time it shocked one of us, the others giggled.

Now we needed some cows to put in this enclosure. Someone handed me the reins to a young, gray horse and told me to go help gather the cows. I swung up on the horse and sat in the old saddle. The saddle had clearly been used to rope many bovines over the years. The saddle was crooked, and I could see that the wooden frame of the saddle had been broken. This had likely happened when someone roped too big of an animal and pulled too hard.

I set out at a trot to join some other cowboys looking for cows and calves. I found an old cowboy on a skinny, bay-colored horse. When I rode up close to him, I realized he only had one functioning eye. The other eye was there but looked like it was dead. Upon closer inspection, I realized his horse also only had one eye. I could hardly believe the irony.

I asked him how I could help. He pointed out into the grassy hills and began mumbling. This guy was just impossible to understand. So, I just rode in the direction he pointed, and sure enough, I found another cowboy and the cows. I helped him drive the cows to our newly-built pasture.

We stopped along the way to let the cows and horses drink some dark, muddy water from wheel ruts that a tractor had made. I noticed pieces of junk attached to some of the cattle. I thought maybe it was not supposed to be there, so I asked the cowboy, and he explained the purpose of the junk.

One cow had an old car license plate attached to her nose. It was lightweight and did not bother her at all, it seemed. I watched her put her head down for a drink. She could drink with no problems. She then reached over and got a mouthful of grass, again without an issue. They said she had this plate to prevent her from trying to drink milk from other cows. The plate made it impossible for her to get a teat in her mouth. We use the same method for solving this problem in the USA, except we use guards made from plastic.

Occasionally, full-grown cows like this one develop the bad habit of suckling from other cows. This causes two issues. The cow is stealing milk that is meant for a little calf. If she steals enough, the calf could starve to death. The second reason is that cows ferment all their food in their rumen.

Bad things can happen when milk enters the rumen and ferments. Young calves have an evolutionary bypass to prevent this issue, a groove in the throat that allows milk to bypass the section of the stomach that ferments. With age, that groove closes. The license plate in the nose prevented big problems for both that cow and calf.

Then, I noticed a cow with a rope tied loosely around her front and back right feet. The rope was made from the strings that hold hay bales together. I thought maybe she had gotten tangled in strings that someone had forgotten to pick up. The cowboy informed me that this rope also served two purposes.

This cow was a terrible mother. She was wild, always trying to run away and leaving her calf behind. The rope on her feet allowed her to walk unimpeded, but she could not run and leave her calf. The cowboys liked this because it made their job easier; they did not have to run her down and put her back in the herd. The second reason for the rope was that when her calf was young, she would kick at him every time he drank. This rope tied to her feet prevented her from harming her calf.

After a long day, we returned to the house. A little fat man was standing outside with a six-shooter in his hand. He was all smiles and excitement when we spoke to him. He was a butcher, I learned. They had a fat pig at the farm, and he was there to kill it. He killed the pig with a well-placed shot. We used a tractor to pick up the pig and set it on a homemade table, made from two oil drums and a pallet.

We built a large fire, and the fat man set the biggest pot I had ever seen on the fire. It was big enough to take a bath in. After filling it with water, we waited for it to boil. We spread an old blanket over the pig and then poured boiling water all over the blanket, steaming the pig underneath.

They set four large, sharp knives on the table. Each man grabbed a knife. They quickly pulled the blanket back and went straight to work. The steam had loosened up the hair follicles, and they removed all the little hairs from the pig. When the pig cooled down, the shaving became difficult. We put the blanket back over the pig and poured more hot water. It was a tedious process. Modern meat processors have machines that do this job.

When the pig was shaved and washed clean, they hung the pig and removed the skin carefully, so that it could be saved to use for leather. The organs were carefully removed, and none of them would be wasted. Then, the meat was cut off piece by piece. These men were artists with a knife: they worked quicker than I had ever seen. When the job was done, the father pulled out his finest bottle of Mezcal, and we all had a drink. We ate like kings that night. It was one of my finest nights in Mexico. Soon after, I headed back to the USA for Christmas.

Mongolia

1
Trip Planning

I had traveled the world all alone. Every single place I visited, I did it by myself. There is something liberating in that, but also something selfish. I often had people ask if they could join me on trips, but I was unwilling to alter my plans to suit them. I knew where I wanted to go and when opportunities arose, I pounced on them.

I would have gone nowhere if I had sat around waiting for people to be ready to go with me. I was lucky to realize that early on and have the strength or stupidity to go on the road alone. Being alone was therapeutic, defining, freeing, terrifying, and sometimes lonely. It sculpted my view of myself and the world in a big way.

Asia was the final continent for me to visit. I planned to travel to Mongolia, and again, there was a small crowd of people behind me saying, "I'd love to go if you'd have me." I gave them the same old response: okay, let's go, here are the dates and the price. As I expected, nearly everyone backed out. The last person remaining was one of my childhood friends. I had known Marley for nearly twenty years at this point. Marley was determined to make it to Mongolia. She had barely ever left the United States before she committed to the trip, a bold move, and one that would pay her back for a lifetime.

Through six months of planning, Marley stuck with me, and all the while I waited for her to come up with some excuse not to go. It was not until we were leaving that I was convinced she was coming with me. We boarded our first flight together from Denver to Los Angeles. It occurred to me I had flown probably fifty times in my life, both domestically and internationally. This was the first time I had ever even been on a flight with someone I knew.

It was a change for me, but a necessary one. This was the first time that I had to think of someone other than myself. I had to consider the opinion of someone else who had given up just as much as me to be on the trip. I also carried a heaviness with me: I wanted to ensure the trip was worthwhile and safe.

Thankfully, I did not have to worry about her abilities. Like me, Marley had spent her whole life on horses. She had an epic job running mule strings in some of the United States' most beautiful national parks. One week before leaving for Mongolia, Marley called to tell me someone had sent her alone with a horse and a mule to do a job. She had to pack the remains of a perished climber from the base of an enormous mountain. My only thought was, "If you can pack a dead body off a mountain all alone, you can surely survive Mongolia." Weirdly, it made me feel safer with the whole situation. I really enjoy people that can handle themselves in tough situations. People often find themselves in tough spots while traveling, especially when traveling with me.

I scoured the internet for information on how to travel Mongolia on horseback. I had gotten the idea in my head that we might purchase two horses, and then just ride across the country: there are no fences and no trespassing, so it just takes time. Then I was approached by a young man, who contacted me through social media about an upcoming trip he was planning to Mongolia. He was trying to visit the last people on earth who still ride reindeer. The Tsaatan people live in northern Mongolia, and are the last semi-nomadic people who depend on reindeer for transport.

The young man needed to make a crew to split costs, or it would not be financially feasible. Based on what he had read and seen about me on social media, he thought I had what he wanted in a team member. We spoke on the phone. I was instantly hesitant about going. If I went with him, he would handle all the logistics. I would just give him a chunk of money and show up. It was completely opposite to anything I had done before.

I spoke with Marley about this. She liked the idea of someone handling all the loose ends. It rubbed me the wrong way, but I decided to hear this guy out. Maybe he could convince me that going with him was better than going at it alone. After we spoke on the phone a few more times, he nearly had me convinced.

With Marley leaning toward the planned trip, I was right in the center. I could go either way. The deciding factor for me was that I had no clue what to expect, buying horses and riding across Mongolia. If I was on my own, it was my risk alone to take, but since I was not, I thought it wise to take the safer route.

I had just barely scraped enough money together for this trip. I could not afford both the horse trip and the reindeer trip. It was certainly a tough choice to make, but in retrospect, not a decision I regret in the least. I wanted to ensure Marley had a good experience.

I had set aside a little over two months of time to stay in Mongolia. Marley only had three weeks to spare. That also gave me over a month alone after she had gone to make my adventure however I wanted, for as cheap as I wanted.

We concluded our planning, and we set our trip dates. My mind started spinning. I became very uneasy about the entire thing for one reason alone: this was the end of the journey. I had spent so many years and so much money on traveling. I had passed up many opportunities to pursue my crazy dream of working as a cowboy on every continent.

What would life be like once I had succeeded? I have heard it said that to be happy, we all need someone to love, someone to look up to, and something to chase. My life would fall apart if I lost that thing I was chasing and could not replace it with some new, lofty aspiration.

2
Getting to Horse Camp

When we arrived in Mongolia, it was not far from what I had expected. Maybe I had just become a little more callused about being dropped into a completely new culture. Maybe after all the turmoil and madness I had experienced across the globe, the shell shock no longer hit me as it did before. The Mongolian airport was small. We found our way through the chaos of the baggage claim and found a taxi driver who had been sent to fetch us.

We climbed in and headed toward our accommodation. The next day, we would meet up with the rest of our crew, with whom we would spend a few weeks in the wilds of Mongolia. I worried a bit if they were people I would be able to put up with for three weeks. It had been years since I had spent that much time with one group of people. I worried about how I would handle it.

When the crew finally assembled, we met over coffee. It went as most first-team meetings go: everyone asks about who you are and what you do, trying to get a feel for each member. It was a menagerie of people from all over the world, with two things in common. We all had horse experience, and we all enjoyed traveling.

Some people rode English, and some rode western. Some had competed in the Mongol Derby, the world's longest horse race. Others had ridden growing up, but it had been some time since they had last ridden a horse. We ate, drank, laughed, and told stories late into the night, becoming more and more familiar.

The trip started with a flight to a small airport in northern Mongolia. We met up with a couple of guys who owned Russian Furgons. A Furgon is a very capable four-wheel-drive van. The van could haul ten people and their gear just about anywhere. We piled in the vans and headed to the border control. We were going to be very close to Russia.

We were entering a protected part of Mongolia, and we needed special permits to enter. We spent the better part of a day outside the permit office, baking in the sun, reading, wandering around the town markets, and eating lunch. The bureaucratic government agency sat on their hands, making our life slightly more difficult. Finally, we got the permits. We had at least eight hours of driving still ahead of us, and the sun was sitting low in the sky.

The vans had small windows, and everyone was packed in tight. The seat in the middle faced backward, and the suspension was rough. All these things coupled together made for some pretty carsick people after eight hours of bouncing down winding, two-track roads. We stopped a few times for people to vomit. People were passing out motion sickness pills like they were some sort of party drug. I never took the pills; I had not been carsick since I was a child. Luckily for me, I never actually got carsick on that journey. I was one of the few.

After our long drive, we arrived in a tiny village where one of our drivers lived. We stayed at his house, piling into a barn for the night. Totally exhausted from the drive, most of us immediately crashed without eating anything. When I woke up early that morning, I went out and watched the sunrise over the little mountains in the distance.

As I watched the orange rays hit the town, the morning fog burned out of sight. I could not wait to be done with the driving leg of the journey. I think everyone felt this way. We drove most of that day as well before arriving at a horse herder camp where we would load up on horses. We stayed with a family that had seven boys. All of them were grown, and most had moved away, leaving Mom with only a few boys to help.

These boys would be our guides up the mountains and to the reindeer herders. One of their brothers had married a girl from a reindeer family, and he now lived there with the family. Everyone was dirty from the car ride, and the temperatures soared that day. The boys took us down to the local river for a swim. We did not know it, but the river was barely above freezing. It was straight off a glacier in the mountains.

We all went bounding into the water before shriveling up and crawling back out just a few seconds later. The boys took turns doing backflips off their horses into the river. Of course, a few wrestling matches broke out as well. These boys were the first country boys I had met in Mongolia. They were covered in scars, dark brown from time in the sun. All of them were lean and fit, incredible horsemen and full of fire, ready to show off for their new guests.

I helped the boys kill a sheep that night. They killed it using a method I had never seen before. First, they flipped the sheep on its back and made an incision in its belly just big enough for one of them to stick their hand in. Then, with the sheep still alive, one boy reached his hand inside and pinched one of the major arteries. Within a few seconds, the sheep had passed out from lack of blood in the brain.

With the animal unconscious, they grabbed the sheep's lungs and collapsed them. It was dead shortly after that. It was like nothing I had ever seen, but it worked quickly. Judging by how the sheep behaved, it was also relatively painless. We cut up the sheep for that night's dinner.

The boys made a small pile of stones and then built a fire around it. The stones stayed in the fire for nearly an hour. Next, the boys cut up potatoes, onions, veggies, and meat, and chucked it all in a large, old metal milk container. Then, they threw the hot rocks from the fire into the container with all the other food and sealed the top. Finally, they set the full container on top of the fire, where it cooked for nearly an hour.

Every ten minutes, they pulled the can off the fire and rolled it across the ground. They did this to stir everything inside before returning it to the fire. When they dumped that can out on a big metal plate, out poured all the veggies and meat and juices like something out of a crock-pot. We ate by firelight like cavemen tearing at the meat with only our hands. It was delicious.

3
Pack Horses and Runaways

The next morning, we woke early, full of anticipation of what was to come. We planned to take off on the horses and head to the mountains. They assigned us our horses, animals they hoped would match our personalities. For the most part, each horse did match its rider's personality. I swung up onto my mount, my first time on a Mongolian horse or saddle. I felt so incredibly uncomfortable and awkward during those first few moments. The saddle was just a metal pack frame with a leather pad laid on top to sit on.

The stirrups were held on by thin nylon straps, nothing like the stiff leather of the Western saddles I was used to. A few people on the trip had competed in the Mongol Derby. One of them was an Australian girl. The Mongolian men wanted me to take a dark-colored horse, but just as I was about to take the reins, they switched us, giving the Australian girl that horse and giving me something else.

Both of us mounted our horses. We were the first ones to get on. We both had a little runaway. Her horse was harder to control than mine. She was nearly out of sight on her horse before she got it turned around. I felt bad for her, getting a semi-crazy horse like that. Then I reminded myself that this was the Mongolian steppe.

I was just trying to survive this ordeal, so it was better her than me on that wild creature. My horse was wild enough as he was. By the end of the trek, the two horses we had ended up being two of the best in the entire group. Neither of them ever got tired, and they always wanted to walk in the front. They were always responsive and willing to move. Most of the other travelers did not get horses like this to ride.

As everyone mounted their horses for the first time, things got even crazier than I expected. First, one person came off, breaking their thumb. Then, the biggest and strongest of all the Mongolian boys hurt his knee badly while trying to trip down an unruly packhorse. Packhorses scattered in every direction, bucking, biting, striking, and running. They had to have hobbles put on them to make them handleable again.

Falling packhorses smashed more than one person's bags. At that moment, I was a little worried we would never make it there, and if we did, that we would make it with only half of the people because the other half had gotten injured. This was my first experience with Mongolian horses. As it turned out, this was just totally normal behavior for them, and everything would be fine.

We finally left horse camp as a big thunderstorm rolled in, soaking us all. Well, it certainly soaked me because I had brought old, second-hand clothing. Most of the other people carried brand new, purpose-built equipment for this trip. I was a little jealous that I did not have nice rain gear.

We wandered down a massive valley where we came across a herd of about fifty camels. A friend and I decided we had to go closer to see them. We took a few pictures. Then, we felt the need to chase them and run with them. Hopefully, the owner was not looking down from a mountaintop somewhere.

We reached a large river crossing late in the afternoon. Thankfully, the river was only shoulder deep at its deepest. This river was the boundary of the flat grass steppe lands we had come from. We were now entering the mountains of northern Mongolia. After crossing, we followed the river, and bugs began swarming my horse. My horse started crow hopping, one random jump at a time. I had the river far below to my left and a cliff to my right. I thought surely I would die there. One boy rode up next to me and smacked my horse hard on the ass. I thought he was trying to make him buck more, but as it turned out, there was a horsefly I had not seen on the horse's ass that was biting him, causing him to buck. I could only give the brother a smile to thank him.

We followed this river until we made camp that night. I think it was the most bug-infested camp I have ever stayed in, but luckily the bugs were mostly flies, not mosquitos. They were still annoying but not did not leave us covered in bug bites. After that, I visited my horse where he was tied up.

I poured some of my bug spray on him where the flies were bothering him most and gave him a scratch under the chin. I named him Chinggis Khaan, after the Mongolian conqueror. He had three white stripes down his belly on each side, scars from his past adventures. I told everyone they were one stripe for each man he had killed.

The boys called me over to their campfire. They offered me some Airag, fermented mares' milk with a low alcohol concentration. They had brought it in a big used soda bottle. The boys used a dirty old tuna can to pass around drinks. When I drank my first tuna can, I knew something was different. They had poured a bunch of vodka into this Airag. The stuff was potent. We finished that bottle around their fire. They had made a second fire upwind from us, burning only dried horse manure to drive the flies away.

I slept well that night beside the creek. When I woke up the next morning, I boiled some water so everyone could have coffee. I would wake up first most mornings. Mongolia seems to be slow getting going in the mornings. If I needed my coffee early, I would have to take matters into my own hands. It was my favorite time of the day, the quiet before everyone else was up.

I watched the sun come up and burn the fog out of the valleys. The pack horses threw the same wild tantrums when we packed up again and prepared to move camp, trying to hop and spin their bodies in any way possible to get away from their handlers or the pack saddle. This time, though, we were ready for it. I jumped in to help hold these horses, as did a few others. Usually, an extra set of hands never hurt in a situation like this.

4

Finding the Tsaatan Herders

We rode higher and higher into the mountains. We would always make camp near the river to ensure easy water for the horses and for cooking and drinking. The weather was like the late summer in Wyoming: warm, with small rainstorms almost daily.

Riding at the bottoms of valleys or on the sides of mountains severely restricted the distance you could see. On the flat ground, people could see storms coming from far away and prepare. On the mountain, we did not know until it was right on top of us. We rode through the pounding rain more than once. It hardly weakened the spirit of everyone there. Visions of the last reindeer riders filled our thoughts.

We finally came to one last very steep push up a mountain that would take us above the treeline. I was fortunate to have one of the fittest, fastest horses of the bunch. This allowed me to stay ahead of the rest of the group, and afforded me the silence that was often lacking. It also offered me seclusion with the Mongolian guides. I got some great photo opportunities by being apart from the pack. By this time, I was taking my photography seriously. I was always on the lookout for the next excellent picture, thinking I might be able to market some shots to a magazine or gallery.

We reached one last big clearing overlooking the valley. It had taken us ages to climb up. To the right was a magnificent stone face almost a thousand feet high. At the top of the pass was a Buddhist shrine. I circled it three times clockwise on my horse to ensure safe passage for me in the valley of the reindeer. One by one, the rest of the group slowly popped up over the top of the pass. Everyone stopped to rest their horses and tighten saddles. After the rest, we followed one boy down the other side. They instructed us to look out for reindeer from here on out. We were now searching for the family; we knew they were in this valley but not exactly where.

We saw our first reindeer about twenty minutes later, but they were high on a hillside above us, and difficult to see. After walking for a while longer, we found the camp. We had to jump our horses over a deep washout as we approached. This made the horses and people nervous, but they knew that once they crossed, they were at the finish line. Anxiousness could not hold anyone back now.

The children all laughed and pointed at us. Domesticated wolves were running around their houses, which looked exactly like the Native American teepee. We tied up our horses, quickly grabbed everything, and went to meet the family. They welcomed us with reindeer milk, cheese, and salt tea, which was also splashed with a heavy dose of reindeer milk. Grandpa was a tiny man, browned from the sun with wrinkles that made him look 20 years older. He smiled a nearly toothless grin when he met me.

Most Mongolian men got a kick out of me immediately because of my beard. By this point, I had a beard long enough to grab with a fist, with some still reaching out the bottom. I must have had one of the largest beards in Mongolia because Mongolian men grow little body hair. Often, old men would draw a line across their throats and point at me. I thought at first that it meant they wanted to kill me. I found out that they were simply saying they wanted to trade heads with me so they could have a large beard.

Just after we arrived, the surrounding families began rounding up all the reindeer. They put all of them in a corral so the families could catch them and tie them up at home for the night. Almost all the reindeer were halter broke. They were like pets, always trying to lick the salt off your skin. I ran to the corral. I had not seen a reindeer up close yet.

I began snapping photos left and right of the little children helping their parents lead the reindeer home. I laid my hands on my first reindeer. They were so much softer than a horse and much more sensitive to touch. Their antlers were large but still in velvet. I found out quickly that velvet is incredibly sensitive. They could feel it even if you blew too hard on them. I imagined this was so they did not hit and damage them while they were still growing so quickly.

Once the reindeer were caught, it was time to lead them home. They would be tied to a stake in the ground for the night. They handed me about four reindeer to lead back to the teepee. It was only then that I realized when reindeer walk, their feet click ever so quietly with every step. I got away from the people talking and the wind blowing, and the air became silent. The clicking of a herd of reindeer feet was a magnificent sound. I had never heard anything like it before.

That night we played with the children in the little encampment of teepees. All the children wanted to touch my beard: they had never seen one so big. They climbed all over me like I was a playground. The brothers were feeling rowdy. They always wanted to practice Mongolia's favorite sport, wrestling.

We were now quite high compared to where we had started. No trees grew at this elevation. When the sun went down, it got cold quickly. Even though it was the middle of the summer, it was still common to have frost on the ground in the morning. We all made our beds on the ground in the teepees. We filled the stove with wood, hoping to have a good enough fire to stay warm. The family brought us horse blankets and some extra clothes to put on top of our sleeping bags. Most of us did not have heavy enough bags for this type of cold.

Gambat was the brother who had married into the reindeer family. He brought me a green Deel, a traditional Mongolian robe, to cover my bag. I slept with the Deel the entire time we visited them. In fact, it was the only thing that made me comfortable enough to sleep at night in the cold. Sleep did not come easy that night, though. Even with the exhaustion of riding multiple full days to get there, I could not sleep. The cold nipped at my feet. I tried to keep them covered. The thought that I would ride a reindeer the next day made my mind race.

5
Sam the Reindeer

I woke early again as usual. The fire had burned out overnight, and I could see my breath. I had no desire to get out of bed and boil the kettle so everyone else could have coffee. After about thirty five minutes of lying in bed dreading the cold, I finally got the urge to rise, so I got up and did it anyway. Grandpa was outside the teepee making some sort of tea. He had an upset stomach.

Grandpa smiled at me, exposing some brown and green chunks from the tea that were sticking to his remaining teeth. I smiled back and wobbled my way over bog knots down a slick riverbank to collect water for the kettle. I walked back past Grandpa as he drank his tea. I got a feeling that he was drinking a special kind of tea, though I did not know what.

The kettle had just boiled when the neighbor stuck his head out of his teepee and yelled at our family's teepee. Brother Gambat stuck his head out, alarmed, and began scrambling for something. When I looked back at the neighboring teepee, I saw a boy who was maybe two years old dragging a large hunting rifle out of the teepee. The gun had a big scope. The father hurried and saddled his horse before grabbing the gun from the boy. The father galloped off without saying a word.

Gambat followed shortly after. With a gun slung over his shoulder, he swung up on his white stallion, bareback, and galloped after the other man. They had spotted a wolf in the next valley over. It had eaten a baby reindeer the night before. A wolf meant all hands on deck. All the able men in the valley rode out for a chance to kill the wolf. It was a good omen if you could kill a wolf. They were quite valuable as well. We got word later that morning that someone had killed the wolf, though we never saw it or heard much more about it. I wished I could have gone along to give chase.

When our Mongolian translator woke that morning, he spoke to Grandpa, who told him about his tea. Grandpa had not felt well for a few days. His stomach hurt. Grandpa had visited the local Shaman to find out what kind of brew could be made to help him. The Tsaatan people who ride the reindeer are still very shamanistic people. The symbols of shamanism surrounded us while staying in their camps.

The Shaman had told Grandpa to make a tea using a mixture of reindeer and horse shit and drink it until he felt better. For almost a day, poor Grandpa lay there drinking his poop tea, smiling at us as we walked by. He was not feeling any better. Many of the team members had brought medicine to help with this. A few were also professionally-trained nurses and paramedics, but Grandpa refused their help. That afternoon, he finally gave in. He swallowed a little pink pill of some sort, and within an hour, he felt better.

"Shocker!" I thought. "Medicine works," but those thoughts gave me a sour feeling. This tiny culture had been under attack from all sides for generations now. Their culture and language were slowly being enveloped and overrun by modern society. They struggled to cling to what little culture they had left.

This made Tsaatans special, making them worth traveling to the other side of the earth to see. I was left conflicted. I am still conflicted to this day about whether it is the right thing to do to take away the poop tea, to give them the little pink pill that makes everything feel better. I could certainly see the argument for both sides.

After breakfast, it was time to saddle up the reindeer. Like the female horses in Mongolia, female reindeer were not for riding. They were for milking and breeding purposes. We rode only the castrated male reindeer. There was one big bull reindeer that was given to the special guests. Our guide rode it most often. He had created an incredibly unique bond with this reindeer family.

These people took him in as family. He brought them joy, entertainment, technology, education, and income with every returning trip to visit. However, a few other people also got to ride the big bull. He was spectacular to ride because he had the biggest antlers; they wrapped nearly behind my back when I was sitting on him. I was given a few brief opportunities to ride the big bull around camp.

I helped Grandpa and brother Gambat put blankets and saddles on the reindeer. The men knew each reindeer by his horn shape and color markings. I would bring them up one at a time in an assembly line fashion. The saddles were largely handmade to fit reindeer. They were almost the same design as the iron and wood horse saddles we had ridden in on, just smaller. The pack saddles were entirely hand-carved out of wood.

Once we had all the reindeer saddled up, they were each assigned a rider. Assignments were based on the person's weight and temperament. I am not a big man. I stand at 5'9" and weigh 170 pounds in my clothes and boots. The animals could only carry people my size and smaller. They could carry more weight, but only for short distances. The people who were any bigger than me had to stick to horses to get around, other than brief rides around the valley.

They informed me that I could not get on a reindeer like a horse. I could not put my foot in the stirrup and swing on, because it would pull on their backs too hard. To get on, I had to swing over in one fluid motion. Some people had to have someone help them. It took me a few tries to mount up by myself. They gave me one of the biggest, strongest reindeer aside from the bull. I named him Samuel M.F. Jackson based on his attitude.

When Sam took his first steps, I was mesmerized. I imagined this is how many people felt sitting on a horse for the first time, a feeling I had forgotten after twenty years of riding. I could not stop smiling. One person watching said, "You don't have to say anything, cowboy. Your face says it all!"

I was floating on a cloud. The animal was much smoother than a horse, and his feet clicked with each step. It was odd to hear that noise. I watched as Marley and a few other westerners were helped onto their reindeer for the first time. They could only make childlike noises to express their feelings, much like me.

Everyone had a hold of their reindeer. Brother Gambat threw his three-year-old son on his own reindeer. Grandpa mounted his reindeer and yelled to us, "Yah way!" meaning "Let's go" in Mongolian as we followed him out of camp. He led five or six animals with just pack saddles on them. One of them carried a chainsaw and gasoline.

We moved out of camp at the speed of molasses on a cold winter day. We were completely out of wood, so we planned to ride out for firewood that day. We needed it for warmth and cooking, come hell or high water, we would bring it back. That clear, sunny morning sky gave no sign of the wrath it would bring that afternoon.

6
Firewood

Grandpa led us through a deep, soft bog that the horses would have sunk in quickly. Instead, the reindeer's large, cloven feet floated us across as if the ground was solid. The horses took the long way around. We wound up through a mountain pass and then down the other side, crossing small streams of water. There were no trees or wood high in the mountain valleys where the families made summer camp. That meant riding a fair distance down the mountain to find large trees to chop down.

We stopped to check that everyone's saddles were okay after crossing a larger stream. There was something eerie about a rock that we had passed. All the reindeer had walked by it cautiously. When one man on a horse in the back of the line passed the rock, his horse, for some reason, spooked and threw him off. The horse then dragged him for some distance before he hit a rock, which yanked his foot out of the saddle.

The horse carried on, galloping and bucking in circles Everyone quickly jumped off their reindeer to prevent more issues. It was quiet and tense for a few moments. Nobody knew what had happened or if everyone was okay. We were too spread out to see much. Finally, we got word that everyone was okay and nobody was going to die, so we carried on out of the valley to find firewood.

I will never know what it was about that big rock that made the animals so jumpy. Both horses and reindeer alike disliked the rock. They spoke of how that valley was full of spirits. Maybe something was happening there that the animals felt. The valley was also full of wolves, which hunted and killed domestic animals regularly. Maybe the animals smelled something wolf-related there.

Grandpa showed us some grass that had an edible base. It was incredibly refreshing and tasty. As we descended the mountain, he sang us songs about animals that he loved. The boys came along with a packhorse and showed us pine needles that you could chew like chewing tobacco, spitting the juices as they formed in your mouth.

The family also showed us which trees produced the best sap, which you could then put in your mouth and chew quickly. It turned itself into pine-flavored chewing gum within ten seconds. These people spent their entire lives surrounded by nature. They were so in tune with their surroundings and full of old knowledge. But, sadly, it seemed like they would lose these parts of their culture in future generations.

Around lunchtime, we reached a spot where Grandpa wanted to cut down a tree. We had not packed a lunch, so tree sap gum would have to keep me going until we got back. We cut down, limbed, and then broke down three different trees into packable pieces in just a few hours. The reindeer enjoyed the break.

Grandpa told us not to tie the reindeer to the trees but to the ground, but we had nothing on the ground to tie them to. Smiling, he poked his fingers down into the soft soil. He dug up a thick tree root in seconds, then ran the rope underneath and secured the animal. When riding the reindeer, the ground seemed normal. When I got off and walked around, I realized that the forest floor was like a super soft mattress. A layer of very thick moss had grown over the rocks and roots, giving them a smooth appearance. It was like walking through heavy snow.

We packed all the trees on the reindeer and headed back up the mountain. A long string of heavily-loaded reindeer pointed back toward camp. The reindeer knew their way home. As we approached, the clouds grew dark. We crossed over the last pass into the valley, and lightning bolts began smashing around us. Freezing rain came at us sideways.

The lighting got too close for comfort. I could hear the whimpers from people behind me, trying to get their rain gear on as quickly as possible. It was cold, wet, and probably dangerous. It sucked. All I could do was laugh, for something told me that moments like this were the ones I would miss when the entire trip was over. I needed to breathe and soak in the energy of what was happening around me.

I went into my own world for a short time. I felt the icy drizzle down my neck. The cold, wet wind carried movement and life through the mountains. The air felt alive with magnetized ions. Energy built and collapsed leading up to each lightning strike and thunderclap. I would feel these feelings again, I knew. But, I would likely never feel them again in the northern Mongolian mountains while sitting on a reindeer.

If only you could catch feelings like that in a jar, you could open it up on occasion to remind yourself on a rough day how bliss feels. Better yet, maybe you could share with a friend who could not be there. If you could sell that feeling to those cold and timid souls who stick to the shadows, you would be the richest man on earth.

Someone called my name when we came close to the teepees. When I came out of my own little world, I looked over at everyone shivering on their animals. Some had tried to stay content, and some people were downright miserable. Finally, someone asked, "You're a real sicko, aren't you? I bet this is your favorite part of the trip. You seem to enjoy this far too much." How right they were. This was the part of reindeer land I will never forget. This was the end of my journey to every continent.

I had found my happy place in my own sick, demented little way. I met a young farmer in Africa once. As my friend introduced us, he told this farmer a brief summary of the places I had been and the work I had done. The farmer looked at me and laughed, asking, "You're just a sucker for punishment, aren't you?" It was not until years later that I realized he was right.

7
Uno

We filled those remaining days at reindeer camp with lots of riding, card games, and learning to make bread, cheese, and traditional food. I would spend my days trying to get incredible photos while reminding myself to be present on the journey. The more I dove into photography, the more I worried I was seeing the world through my camera lens and not my own eyes.

My mind was constantly wondering where the next great shot was. Where can I be that nobody else is? I was developing what people refer to as a photographer's eye. I was proud of that. Unfortunately, a photographer's eye means knowing what everyone else likes. The art in photography is certainly in finding your own style and staying true to it. I began passing up some great photos so I could just be there in the present. I wanted to fully enjoy moments without my camera and not worry about the opinions of others. I highly recommend this to any photographer that does not already do it.

Our only bathing was in the ice-cold stream that passed by the camp. The bath was anywhere downstream from camp that a person felt comfortable getting naked. The girls usually walked farther downstream and out of sight from the boys.

This was my first time traveling with a translator. It was incredible the difference a translator made, and what an entirely new level of connection I was able to develop with the people around me. One cool, clear evening, once we were comfortable with each other, our Mongolian mother had us all do a short but powerful guided meditation where we all went to our happy place. She then had us share the happiest place we could imagine. The tears came pouring out for many in the teepee. She took us to a deep and pure place that would have never been reached without a great translator.

Card games were another way of connecting deeply and spending hours together. Once everyone knew the rules, it did not matter what language you spoke. Uno became our favorite game. The children in the camp picked it up so quickly. I would sit down with two or three people on the ground, and we would play a round of Uno.

Before I knew it, I had three kids sitting on me. Both my legs had fallen asleep from their weight. A little one was pulling my beard, wiping snot on my shirt while he made faces at me, distracting me from the game. This would happen daily, multiple times per day even. It was one of my favorite experiences of the entire trip.

Another activity the boys loved was wrestling, Mongolian style. The rules were simple: get the other guy on the ground however you want. When anything but his feet touches the ground, the match is over. The brothers were all incredibly fit young men, lean and strong, and I could never beat them. If one of the boys walked up and grabbed both of my wrists, I could respond with "no, no, no" and submissively back out. They considered anything else the first move of the match. I had to be ready to wrestle at all times.

One morning when I was feeling froggy, brother Gambat walked up to me, grabbing both wrists, and we went at it. The pulling, pushing, and twisting lasted maybe a minute or two, but it seemed like an hour. In the end, by the skin of my teeth, I threw him down, my first and only Mongolian victory. I earned some respect from him. He stood quickly, demanding a rematch, which I almost won, barely tipping as he had done.

Regarding effort, Mongolia does not differ from anywhere else in the world. Effort transcends language and cultural lines. The willingness to try someone else's culture goes farther than anything I have ever seen. I have heard it said many ways in cowboy culture. A man's "try" will earn him respect quicker than anything else.

When people see you giving it your all, that is when they help, teach, and invite you into their world. Wrestling, cards, horsemanship, and respect carried me so far in Mongolia. I do not think this is an international thing. I think this is an interhuman thing. You could be in your hometown, in your own backyard, and your try will be noticed and respected by those around you.

While there, I traded and haggled for some cool handmade knives and other items carved from bone. I will keep all of these things for the rest of my life, but there ended up being one item more special than anything else. The morning I was leaving, I asked Gambat if he would sell me the old Mongolian robe I had used as an extra blanket. It was torn, burned, and dirty, so when I asked, he was surprised.

Gambat told me I could just have it. But, I could not just take it. I wanted to trade him my coat for it. I could not live with myself knowing that he would be down one coat living four days' ride from town and having as little as he did. But, the simple gesture that he just wanted me to have something of his was something

I will never forget. I also left him with a Leatherman knife for all the kind things he had done for me.

We rode out of that valley, waving goodbye to the reindeer and the family that had taken us into their home. I wondered if I would ever be back there again. I worried how long their culture would last there on the Mongolian-Russian border. It was a quiet day on the trail. Nobody talked as they had riding up into the mountains. I could feel a little sadness that everyone was emitting being carried around by the wind.

The horse boys did what they could to keep spirits high. They loved to sing, and they were exceptionally good at it. They would sing at the top of their lungs for hours, songs about Mongolian animals and traditions. When I look back at the videos I took, I am amazed by the singing echoing through the valleys. No matter where I am, it brings me into some sort of primal, dreamlike state. I could hide there all day.

8
Back to Civilization

The horses quickened their steps. They knew we had turned to head home now, and they were walking downhill. We hugged a large, fast-flowing river almost all the way to the bottom of the valley. Then, we split off from the river to head for a low mountain pass. I knew where we were now. My mental compass was working correctly.

Just on the other side of this ridge, we would drop down onto the flat, treeless grasslands of the steppe. Then, a thunderstorm rolled in on top of us, dumping rain on us for a short time, an occurrence I had become quite used to. This was Mongolia's peak rainy season. It rained almost daily.

We set up camp in the valley and crossed onto the steppe the next day. When we stopped to inspect the ground and water, the pack horses put their heads down to eat. When one horse pulled its head back up, he jerked his lead rope under another horse's tail. Like many young horses, the horse did not like this feeling, and it caused him to instantly react and then buck. Luckily, this horse was being ridden by the wildest brother of them all, Gansuk.

Gansuk was the youngest, and he could ride anything. His horse crashed through tall bushes, through the mud, over small streams, then through the rest of the horses. Gansuk could do nothing but smile and laugh. His horse bucked, jumped, kicked, snorted, and ran until he had enough. Gansuk stepped off, still laughing, and pulled the rope from his tail. He then tied his horse up and made his camp for the night.

The way the Mongolian boys made camp was incredible. They traveled so light and made whatever they did not have. Their tent was just a large canvas that doubled as a thick saddle pad on the pack horses during the day. They carried no tent poles. Two pieces of wood were all they needed to make a frame. They carried no tent stakes, so, with a hatchet, they quickly carved out ten or more when they got to camp. All said and done, two men could create everything except the canvas in less than ten minutes.

These men often traveled with one big, metal bowl to cook with. They made pine needle tea in the mornings. Their wives or mothers often prepared them some noodles and dried meat to eat. They also carried awfully hard, dry bread; it kept well in the saddlebags for days. If they were lucky, they may get some fresh meat or shoot something along the way that they could eat. They brought only the clothes they were wearing, and many of them had no pad or blanket to sleep with. They just slept in their clothes, wearing the big, heavy robe they called a Deel. It kept them nice and warm.

The next morning, we packed up our camp for the last time. Only one obstacle was left in our way, the big river. We had crossed it on our way out, and it had been fairly deep. It had been raining every day, and the river had swelled. We scoured the bank for a suitable spot before committing to one place. I pulled my feet high on my horse's neck to keep them dry. I worried about all my expensive camera equipment on the pack horses getting soaked.

Many people just kept their feet in the stirrups. This was maybe the safer option, but they certainly filled their boots, while mine stayed dry. We had to stop and check everything when we reached the other side. The brother's uncle, who had been riding with us the entire time, galloped away from the group. Five minutes later, he came back. He rode up to where I lay in the grass, resting, letting my horse graze. He had something for me, he said, a gift. We had become quite close during the trip. Even though we did not speak each other's language, we understood each other well. I had given him a pocketknife to thank him for teaching me everything he had.

Uncle held out his closed hand with his gift for me. When he opened his hand, it nearly brought me to tears. It was a wolf's ankle bone. These bones are rare. It is a tremendous honor to receive one. He tied it to my belt, telling me that this bone would protect me in my travels. It would give me the power and the bravery of the wolf in all challenges that came my way. I tapped my hand on my forehead a few times. This was a way to show thanks and respect in Mongolia. To this day, the wolf bone is one of my most prized possessions. I take it with me everywhere I go.

We stopped at a village in the middle of nowhere. Everyone started tying their horses to the fence to go inside. Some people did not know that you must tie your horse to the post, not the rail. One horse panicked, and thankfully no people were close by. When he panicked, he pulled the fence rail off, scaring more horses that also began to panic.

Within seconds, we had removed a large section of this guest house's fence. Four of our horses were loose and galloping away with the fence dragging behind. One brother caught the runaway horses. I looked at my trusty horse, "Chinggis Khaan." He still stood calmly by the fence. When I first got on him, he was one of the craziest horses in the group. He never really changed. I just learned how to roll with him. I am so glad I did.

The group was done with horses now. We threw our bags into the backs of the Russian Furgons again. We bounced for ten hours back to a paved road, and drove some more before arriving at the tiny airport where we had started. Our flight had been delayed for five hours, so we just sat around the tiny airport. It had been exceptional, leaving the world and our phones behind. Nobody had pulled their phone out in so long. We had been without a signal most of the time.

Now that we were back in civilization, everyone buried their faces in technology again. It sucked us back into our everyday lives. We headed back to Ulaanbaatar that afternoon. Everyone split up and said their goodbyes. Most of the crew had adult jobs back home. They would fly home the next day and return to the grind. But not me. I still had over a month left in Mongolia. One adventure was ending, but another was just beginning.

9
Training Trekkers

After returning to Mongolia's capital city, I got a job working for a place that sold horses to tourists. They hired me as a horseman. They needed someone who could ride a horse well and speak English with the clients. Volunteers from all over the world ran the place. I met people from Sweden, Switzerland, Israel, Denmark, France, Norway, Canada, Spain, Russia, Chile, Germany, Korea, and many more.

I spent my mornings gathering the horses from the sides of the surrounding hills so the tourists could ride them. I did this with the Mongolian horsemen. It is astounding to me still how much you can learn from people who cannot speak to you.

Some mornings, I covered ten miles before breakfast, running up and down ridgelines looking for the horses in the fog. I had no option but to ride more like a Mongolian when I covered many miles daily. Mongolia has no fences. I imagine that is like what the western United States used to be like. So, depending on the horses and the weather, on some mornings we had to ride much farther than others.

Gathering the horses quickly became one of my favorite parts of the job. It was also the most testing of my patience. There is truly no other feeling in the world like flying at a gallop behind a herd of galloping horses. Their hooves pound like a quick-moving thunderstorm rolling off the mountain. The mares and foals call out to each other between sharp breaths of air, trying to keep their massive lungs full. The stallion somehow finds his way to the back, keeping a watchful eye on the shape of his herd, ready to strike out at some unassuming young male horse trying to sneak in amongst his girls.

Mongolians seem more at home on horseback than on their own two feet. It is something in their blood. I have seen taxi drivers deliver new clients, then jump on a horse and outride me. Then, they jump back in their car and head back to the city to pick up more tourists.

Mongolians ride in a very jockey-like style, stirrups short, and they stand up higher as the horse goes faster. When the horse reaches a full gallop, they stand straight up. This allows them to use their long polls to catch wild horses. If they do not use a poll, they can swing a big heavy rope resembling a reata and catch horses that way.

As time went on, I tried to ride more like them, standing when I felt like I could balance. I tied my stirrups together underneath the horse, so I could hang off the side of the horse and grab things. The Mongolian riding style requires a lot of confidence at first. I needed to trust my horse and my reflexes because I created so much distance between myself and the horse by standing. It felt like being a stunt rider, almost. When I created this distance, I gave myself little time to react if the horse changed direction.

The other fun part of this job was that I helped train the riders planning to leave on solo treks. Solo trekkers showed up at our company to do a short training course. Then they headed out into the Mongolian steppe to travel alone for weeks, with their newly purchased saddle and pack horses. I loved helping these people out.

I had to teach them all the knots, how to pack a horse, horse health, and common sense with horses. Some even had to be taught how to ride. I taught them how to navigate with maps, communicate on a basic level, and how to respect Mongolian traditions when locals invited them into their homes. I knew some of these things when I arrived there, and the rest I learned in Mongolia with the reindeer. I could teach them. It was an invaluable learning experience for me.

I also learned heaps from the people's solo trek training. This had been my original plan in Mongolia before I paid to go to the reindeer tribe. Unfortunately, the trip to the tribe had used up my entire budget, so I no longer had the money to buy horses and ride out alone. Every single person who showed up to solo trek had less experience than me with horses. Their lack of experience made these people incredible to me.

Trekkers showed up to learn and take on a monumental and dangerous task. These are the type of people I enjoy being around. They were all positive, upbeat, and ready to take on whatever you threw at them. In today's world, I think people like this have become increasingly rare.

After repeatedly doing the horse training for these green solo trekkers, I learned cool new knots. I became good at packing the horses. I knew how to take care of a horse on the steppe. It made me so desperately want to buy my own horses and leave. I loved when the solo trekkers returned. Through hardship, they would develop genius ways of tweaking my strategies or create new ones that were all around better.

One Australian professional raft guide returned to camp after a month. He showed me an entirely new way to pack a horse that he learned from some Mongolian on the trail. He then put his own little tweaks on the knots that he used on his boats. The secret to his way of packing was using an old feed sack on the bottom of his pack saddle as a girth so it did not rub the horse.

This job was not exactly what I wanted, but it enabled me to stay in Mongolia for a month longer for free. I could not pass that up. During my travels, I avoid tourist-focused places as much as possible. I will always go live and work like the locals if I have the option. I still learned a lot. More importantly, I met some incredible people from all over the globe. Some of them became my greatest teachers and friends in Mongolia.

10
Boina

Every Mongolian I met thoroughly impressed me with their riding ability. However, there was one man that stood above the rest. One day, a small Mongolian man showed up around lunchtime. He marched right up to me and, in very broken English, asked, "You are cowboy, no?" and gave me a huge smile. He shook my hand and introduced himself as Boina.

Boina is a nickname. I do not know his real name, but they explained to me that Boina means good energy or the opposite of evil energy. He often tamed wild horses to ride for the first time. Standing behind a nervous mare to milk her in a vulnerable position, he would whisper to them, "Boina, boina, boina." These words would somehow leave him unscathed.

Boina had been called in to help us when we were shorthanded. He had worked for the company for years and now returned when they needed a little extra help. He told me he once worked with another American cowboy. This man had apparently brought his

American saddle with him. They rode colts together somewhere. This gave Boina the impression that all cowboys were keen to jump on anything with four legs.

I could see a minor disappointment in Boina's eyes when he realized I was not just going to jump on any wild horse we caught. Maybe there was a time in my life when I thought this was enjoyable. The broken bones and concussions from past wrecks were a stark reminder. A person can still have a lot of fun and learn quite a bit from convincing some other poor soul with a bigger ego to try it instead.

The first day I worked with Boina, I impressed him with my ability to rope horses in the round corral. He enjoyed watching me do it so much that he kept telling me to rope more horses. When I asked, "Which horses do we need?" he responded, "Rope any, rope any, it's okay," and chuckled. I knew there were some wild horses in the corral that had never been roped, but I did not know which ones. I sailed a big loop at a skinny-looking, flaxen-maned gelding I had ridden before. He wasn't wild, I knew that.

As the rope flew toward his head, he pulled the oldest trick in the book. He ducked the loop and let it fly right by his head and land squarely around the neck of the biggest mare in the pen. This mare had never been touched before. She bolted immediately, and I sat down and began getting pulled around the corral through all the other horses. A cloud of dust followed as I left a big trail of denim and skin on the ground where she had dragged me.

Boina watched with a smile. Finally, he had seen the cowboy take on a wild horse like he wanted. He finished his cigarette and flicked it away before climbing into the corral to help me. This mare fought us for nearly half an hour as we tried to get close enough to pull the rope off her neck. Boina laughed the entire time. It seemed he was most at home when out-of-control renegade horses were involved.

Boina finally got close to her head. All three of us were dripping with sweat now from the fight, and she struck out with both front feet, trying to hit Boina. He whispered to her, "Boina, boina, boina," as he tried to pull the rope off. She again stood on her back feet, striking at us, and he yelled, "PULL!" and we gave one giant heave. The mare tipped on her side in a dust cloud, and we dove on top of her. In a pile of horse and man-flesh, we kept her on the ground.

I sat on her neck, whispering to her, apologizing quietly. I had not wanted to catch her. I did not want this fight, but fate had brought it on us. I looked in her wild eye. She breathed hard, just like we did. I covered her eye so our movement above her would not frighten her.

We slowly took the rope off her neck, and when it was clear, I jumped off her, letting her up. Both man and horse stood up from the ground, a little wiser and with a bit more respect for the other. When things got dangerous and scary, I had not backed down, which is what Boina wanted to see.

It seems that the best friendships are forged through dangerous and scary times; that is when your medals are earned. Unfortunately, some people rarely, if ever, enter this realm. It is important for me to touch the void every once in a while, giving fear and death a little nudge, to let them know I am not done yet. People who avoid that realm lead long, bland, flavorless lives that seem to drag on forever.

People who avoid this realm, this feeling, will never befriend a man like Boina. A man like Boina only has one type of friend, the type that is there when shit gets scary. When you live as he does, always trying to touch the void, you do not have time for timid people waiting in the shadows for the dust to settle.

Boina would always rope the freshest, fattest horses. If they would not buck, then he did not want them. So, he would rope a colt in the morning, walk down the rope, and ear twitch it to take the rope off. When the rope was off, he would let the ear go and, in one motion, swing up on the horse and grab a handful of mane as the horse hogged and bucked through fifty other horses in the corral.

Boina would pull off his hat and fan the horse until the horse gave in and stopped bucking. Then, after jumping off, he would pick up the rope and rope his horse for the day. He did not need coffee in the morning. He just needed a little adrenaline, a little excitement between his legs. So, he would break one horse before catching a much bigger, stronger, wilder horse that he would ride for the rest of the day.

Boina and I took some guests out in a terrible rain and thunderstorm one afternoon. The guests were excellent riders and ready to get home, so we decided we would run the horses most of the way home. I was riding my favorite horse, a half-wild gelding named Lauren who loved to run. Lauren would prance sideways for hours, waiting for his chance to run.

I was told the word "Lauren" in Mongolian meant the opposite of calm and quiet. Maybe something like an adrenaline rush. It fit my horse well. We reached a massive flat area and decided we would have a race. I thought there was no way anyone would beat Lauren and me. It was pouring rain. The ground was slick, and my horse and I flew, leaving everyone behind us.

We approached the end of the flat area and ran uphill. I heard Boina's little chuckle in my right ear and felt someone touch my right shoulder. Boina was standing up in the seat of his saddle like a stunt rider. We were running at full gallop. In the rain on the side of a slippery hill, he stood up on his horse while smoking a cigarette.

Boina had no shirt on and beat me in our race at the last second. He landed backward when he dropped back down onto his horse's back, still smoking his cigarette, giggling, letting go of his reins, and letting the horse go where it wanted. The horse he was riding had only just had a saddle put on it that morning for the first time.

Boina became a good friend and teacher. He was still only twenty-seven years old. He was possibly the most reckless but incredible horseman I have ever seen. He told me he trained racehorses in the Gobi Desert. In the winter, he moved to a horse camp in the north, where he would take care of 500 horses all by himself. There are very few trees in Mongolia, so in the cold winters, there is no wood to burn.

Boina told me he gave up on having fires in the winter long ago. Instead, he stayed warm by drinking Russian vodka and eating horse meat; he needed nothing else. He would catch one horse and ride it through the deep snow for days, herding the other horses. When his saddle horse became too tired, he just caught another horse from the herd and threw his saddle on it. Then, he took the life of the horse he had been riding so he could eat it. Horsemeat is one of Mongolians' favorite things in the winter. They say it keeps them warm. Skinny and sick horses do not survive the winter, so they eat them before they starve.

11
Mare Milking

Mongolians' diets largely consist of dairy products in the summer. This is because they live off what their animals give them, and animals only produce milk in the summer. If they can live off only dairy and bread all summer, that saves them from having to butcher their animals. Then, they can save them for winter meat. One of the most famous drinks in Mongolia is called Airag. It is fermented mares' milk that has about five percent alcohol when finished. Each family has a slightly different procedure for making it. Once fermented, it can be stored for much longer than normal milk.

I drank plenty of Airag when I was in Mongolia. It just tastes like very acidic, sour milk. I eventually got used to it, though most foreigners tried it once and never again. The Mongolian families loved to share it with me. It was rude to refuse. I would often get a belly full of it, then have to ride my horse back home, trotting with a shaking belly full of fermented milk the whole way. It never made me sick somehow. It gave me guts like a dog.

To make the Airag, they had to milk the mares multiple times daily. Milking was all done by hand. Usually, it was the wife's job to milk and the husband's job to hold the horses. The young foals were caught and tied to a picket line. If the foals could not walk away, the mothers would always stay close. When the horses got used to being milked after a few weeks, they were much like cows and could easily be handled.

We were tasked with getting our entire herd of mares milking while I was there. That meant capturing the foals and handling them for the first time. Their mothers also needed to get used to being handled and milked. We did it in true Mongolian fashion, with no corral. We gathered over 100 of our horses and had outriders keep them in a tight group.

When the group was held together, one by one, the foals were roped out of the bunch and tied to the line. The foals were about two months old, easily big enough to drag a man around and hurt him. I roped a few foals while standing on foot, as did the other horsemen. We had tourists keep the horses in a tight bunch.

They often roped the foals at a run, so you had to immediately prepare for impact when the loop fell over their necks. We never used gloves. After a day of this, my hands were destroyed by rope burns. My ribs and hip bones were also burned from being dragged by horses over rough ground. As we caught more and more of the foals, we were left with only the biggest and fastest ones that we had not caught yet.

One foal walked up to his mother's side and put its head under her belly to suck. This left him vulnerable. I could walk up and grab his tail. I did so, then I held on for dear life. Luckily, a few Mongolian kids came to help and got a hold of the foal's head. We captured him and tied him to the line with the others. Unfortunately, the knot that one boy had tied broke, and the foal got loose and ran back to his mother.

The colt made the same mistake again of putting his head down. One of the Mongolians grabbed his tail this time and held on just as I had. The foal must have learned from his first mistake. The foal quickly kicked both back feet in the air, smashing the man's face with a sickening thud. He fell to the ground, limp. My heart sank.

I was the closest person to him. I ran to his side and stood over him to ensure the circling horses did not step on him. When I kneeled next to him, a stream of blood squirted from his mouth, coating my hands and arms. He was coming back to consciousness. I felt better after he moved. He began spitting out bloody chunks that looked like teeth. I found a dirty rag and pressed it to the bleeding to stop it.

His eyes looked up at me, empty and confused, a look that reminded me all too much of my bull riding career. I found a jug of water and began washing the wounds. He could not move his jaw, so I could not see his teeth. His top and bottom lip were ripped in half where the hooves had hit. We called a car, and they rushed him to the hospital. The man already had a broken hand from a fist fight the week before that would also need to be fixed.

I found out later that he would be okay. He just needed some stitches to put his face back together. He would return the next day and ride horses again like nothing had happened. I thought about the fact that I had done the same thing as him just moments before and had no issues. He had his face rearranged, and it could have been significantly worse. It was another stark reminder to me of how close to the edge we all live, how dangerous things become normal to us easily, and how we shut out the idea that anything bad could happen.

I sat there washing the blood off my hands and arms. A little Mongolian boy poured water over my hands for me as I washed in a trancelike state. A colt whizzed by, dragging a rope around his neck. The boy bailed on the rope. I snapped back to reality and helped the boy. We wrestled with the foal and tied it to the line, back to work as usual. We had no time to think of danger, fear, or bad outcomes; we had to catch the foals. We had to milk the mares. Families depended on it.

When we had only a few of the foals left, men on horses chased them with long ropes. Mongolians traditionally use a long pole with a loop on the end to catch horses. We did not have them, so we used the old way of the cowboy, galloping through rough country, swinging a big loop above our heads to throw at the foal. One man caught one quickly, and I wondered how he would hold the rope while sitting in the saddle.

There was nowhere to tie off a rope on a Mongolian saddle, or so I thought. He quickly popped up behind his saddle. He sat on his horse's butt while it ran along, keeping pace with the young horse. The man quickly wrapped the rope around his entire saddle twice and slowed his horse. This allowed us to grab the foal and wrestle it to the line with the others. I could not believe he had just used his entire saddle like a saddle horn while sitting on his horse's butt at a gallop.

At last, someone roped the final young horse. When they tried to hold the rope, it burned through their hands, and they lost it. The horse galloped around the herd with the rope dragging behind it. Normally people would have just tried to throw another loop around its neck, but not in Mongolia. A group of young men on horses galloped along, chasing the rope. They tied their stirrups under the horses' bellies, which enabled them to lean far off the horse's side and grab things off the ground.

At full gallop, the men leaned down as rocks whizzed by their heads. Their horses stumbled to keep their feet and support the men's weight. Finally, of course, Boina, the wildest horse man of all, reached down, grabbed the rope's tail, and stopped the runaway foal. We tied him to the line.

Boina had to show off once we were done. The man had more energy than anyone I had ever seen. He was a small man. When he took his shirt off, which was very often, I could see every muscle in his body. He looked like a small, lean cage fighter. He threw a loop over a huge, white, wild stallion. When the stallion pulled, he sat on the ground and wrapped the rope around his waist. The stallion dragged him, hardly realizing the weight, and Boina held tight.

The stallion dragged him almost one hundred meters over the rocky ground before stopping. Boina had ruined his pants and shredded his legs. He worked his way up the rope to the stallion, grabbed his ear, pulled the rope off his neck, and swung a leg over him. Then, holding nothing but the mane, he rode the stallion. The stallion bucked and jumped and swooped. All his movements were made more incredible by his long, white mane and tail. Boina sat there until the stallion gave up. Then, with a little chuckle, he jumped off and patted him on the neck, letting him go.

We caught all the young horses, then caught their mothers and put hobbles on them, a short piece of rope or leather that is tied between two or three of the horse's legs. This allows them to walk but they cannot run or kick. We carefully milked the mares for the first time, and that night we drank fresh mares' milk. I had never tasted such sweet milk in my life. It was wonderful. Of course, the milk tasted better to all of us because we were covered in dust, with torn clothes, and hands burned by ropes. The milk was delicious because we had earned it.

12
Time to Move On

After nearly a month of working at the tourist horse camp, the atmosphere changed. Things boiled over between the international staff and the Mongolian ownership. Understanding someone completely is so important to running a business. The language barrier seemed to prove too much.

International guests felt like they were not being given what they were promised. Slowly, they began leaving the company. The Mongolian owners felt that the long-term international guests were lazy and needy. I was stuck in the middle of all this because I fell between local staff and international guests. Bad energy swirled around the property for weeks.

I took a few days off to go to the city and clear my head. Instantly, after leaving the horse property, all the problems we were having seemed so tiny. By surrounding myself with the problem daily, it had become much bigger in my mind. This was a big learning moment for me. After a few days in the city, I went back with a clear head.

I was determined to finish my stay and make it as good as possible. I was first contacted by the company. They asked if I was willing to help them. In addition to my usual work with the tourists, I was told I would accompany the solo travelers for parts of their journeys, going on long treks on horseback through the wilderness for days on end. For almost a month, I had only taken people on the same, two-hour horse rides. I could make all the loops with my eyes closed by that point.

I found immense joy with the solo trekkers, both teaching and learning from them. All of them took my number as they left on their big adventures. Almost every single one called me on their journey with some sort of problem. They had nobody to call but me in some situations. When they had injured horses or were sick or a little lost, I would get a call.

We built a strong bond before each person left. Then, when they returned weeks later, we would sit down, and they would tell me everything that worked and did not work. We laughed about all the wild things they had encountered. I sat and listened like it was a bedtime story, envious of their freedom, dreaming of getting away as I made the same loop with tourists daily.

Mongolian horses are tough, but I began having moral objections to the treatment of the horses. The company began overusing horses for no apparent reason. I have a bachelor's in veterinary science. I have spent almost two decades around horses. This does not make me an expert, but it makes me more than just a casual observer. Every day, I would voice my opinion that we were saddling horses that were totally exhausted and too skinny. Or, I would complain that we were putting wet, dirty saddle pads on horses with open sores that were clearly visible on their backs. My concerns fell on deaf ears.

I just chalked it up to something lost in communication. Then I spoke with the owner, who spoke English well enough to understand exactly what I was saying. I asked for antibiotics and spray to cover the wounds. I also asked that he speak to the head horseman and tell him to stop overusing the same horses. He did not give me anything to help the horses, but he spoke to the head horseman.

The next day, the head horseman roped out a horse with a sore on both sides of its back, a raw meat area larger than a dollar bill that had hardly even scabbed over from the day before. When I said "NO!" he laughed and threw a saddle pad on it where he knew no guest or owner would ever see the wound. I continued to tell the owners.

The other staff and even the guests began complaining about horses with sores or horses that were too skinny. I just let them. I encouraged them. I thought the owners might hear paying customers if they did not hear my voice. But unfortunately, their complaints seemed to also fall on deaf ears. Things would get better for one or two days and then revert back to normal.

One morning, we caught a young horse who was a handful. His name was Socks, and only guides rode him. The guy who rode him the day before had rubbed a hole in his withers. It had rained all night, and when he came back in, he had a small fountain of puss oozing out of his shoulders. Unbelievably, they caught him to ride. I spoke to one of the horse guides, who also spoke English. I told him that we should let the wound heal. He replied, "No, we will ride him. Putting pressure on it all day will drain the infection and help him heal. It's the Mongolian way."

I thought he had made that up, just so I would shut up. I had come here to learn. Mongolian horses had already surprised me multiple times with their toughness and resilience. So, maybe riding a wounded horse was the Mongolian way, and it would get better. Instead, the next day the wound was twice as bad, with even more infection and blood draining. I desperately wanted to say, "I told you so!" but that would have accomplished nothing.

After that ordeal, I decided I was through fighting with them about horse welfare. After all, they were not my horses. They did not bring me in to tell them how to run their animals. Maybe it was just a cultural thing, but none of the other horses in the area had these massive, draining wounds. I could not see their ribs poking out slightly like some of our horses. I decided it was something I probably could not change. It was not something I would stand for, and I began looking for a new plan, a new place to go.

I had developed a large saddle sore on my leg from spending all day every day on a horse. My leg began draining blood and puss. I knew I would have to stop riding for a bit to let my wound heal. I would not be treated like a horse.

I received a phone call from an Australian guy out on a solo trek named Tim. Tim was tired, his horses were tired, and he was ready to come home. We organized a horse truck, and he returned to horse camp.

Tim was shocked that the company had fallen apart as badly as it had since he had left three weeks ago. Nearly every single international person working there had left. Later that night, over cheap Russian vodka, he invited me to go camping with him. He had all the gear to live in the wilderness. He no longer wanted to pay to stay at this horse camp. We dreamed up a very loose plan. I would leave with him.

I really enjoyed working for this company. I think they were genuinely great people that ran a great business. I must have just entered at the wrong timeframe while they were transitioning management. The horse welfare problems eventually got fixed. The staff problems would be fixed as well. After traveling so far and spending so much money, I just needed to not be surrounded by that terrible energy. I knew my time would be better served elsewhere.

13
Orkhon Valley

Tim was a man with many interesting stories to tell, stories from his time as a river guide. He taught me more knots than anyone else ever has. Tim had grown up in rural Australia on a property with cattle, but you would never guess that by the look of him. He had long hair below his shoulders. His hair was twisting up into dreads in some places. He had a mustache that rounded out his appearance. He commonly referred to himself as a dirtbag.

Dirtbag to some people is a condescending term, but to others, it is a badge of honor, and to Tim, it was the latter. Tim explained to me that to be a dirtbag is to live simply, to adventure at incredibly low cost by willing to be a little dirtier and less comfortable than everyone else. To dirtbag is most simply to throw your sleeping bag in the dirt and sleep there, where it is free, where you are free. It took little convincing to make me see we could go very far this way and quite cheaply.

Tim and I ventured into the country's capital city for a day or two. We stocked up on supplies and stashed the gear we did not need at a friend's place. We did not know where we would go. All we knew was that we had to pay to stay in a building. Staying in a tent was free.

I spent a morning researching places to go in Mongolia. It is an awfully expensive and large country to travel in. I read on a blog somewhere that you could catch a bus from the current capital to the ancient capital, the capital of Chinggis Khan's empire. The bus ticket cost less than a meal at McDonalds. So, the next day, we jumped on that bus and headed to freedom without proper plans.

When we arrived, we found a super cheap camp that helped us find a ride into the wilderness. After a three-hour drive down rough, two-track roads, the taxi driver stopped and pointed north. He said, "Waterfall, three hundred meters," and then got in his car and drove away. He had dropped us in a totally different spot than we had planned. We were now in the middle of Orkhon Valley.

Orkhon Valley is one of the most beautiful areas in Mongolia and a UNESCO World Heritage Site. We had nowhere to go, nowhere to be, and we did not really know where anything was, so we just walked. A small stove, a lean week's worth of food, sleeping bags, and a tent were all we carried on our backs. The air conditioning on our bus ride had made me quite sick, so we could not make good time, but we plugged along.

Eventually, we found a nice camp spot next to one of the many rivers. It was flat and had lots of firewood. We soon found out we were close to the eight lakes region of the valley, which was the first time either of us had much of an idea where we even were on a map. We had just brought a bag of pasta and some cookies for food. Surely, that would hold us for a week.

The next morning, I sat near our small campfire, drinking some freshly-brewed coffee. A group of five riders passed by in the distance on horses. They were much too far away to recognize their faces. Luckily, I have always had a good eye for horses. I jumped up when I realized those were some of the solo trekkers I had helped train. They had ridden nearly two hundred twenty five miles since I had last seen them. They did not know I was going to be in the valley. I began yelling their names and waving my arms. They were shocked.

They trotted over in complete disbelief. Tim and I, filthy and smelling of campfire, splashed our way through the river to meet them. They told us tales of all they had seen. They gave us a better understanding of where we were. We sat there, catching up on the riverbank for almost an hour. All of us were in complete disbelief of the luck it had taken for us to come together in this spot, nearly a four-week ride from where it all started. Two of the girls told us about a little supply store in the next valley over.

The girls asked if we needed anything. I think it was clear how light we traveled. Without hesitation, Tim said, "Well, beer if they have it, of course!" and the girls laughed. Thirty minutes later, the girls came galloping back over the horizon and rode right up to our camp. They pulled four gallons of beer from their saddle-bags. But, they warned, "Don't open them yet. We galloped the whole way. They're shaken up for certain."

The girls gave us a smile, said goodbye, and wheeled their horses around and rode off. Tim picked the beers up and began building a rock pool in the river to keep them cold. He turned to me and said, "I can't believe this. Not only did they run into us here, but then they turned around and brought us cold beer. We could not be any farther out in the middle of nowhere."

The beer lasted us easily for the rest of our trip. Tim and I spent many late hours around the fire, probing the other for the answers to many of life's great questions. There was nothing but firelight in the vast valley. The stars there reminded me of my home in Wyoming, one of the few places I have seen the stars without the pollution of city lights.

Tim and I spent our days along the edges of the rivers. Tim caught a raven there, and we befriended him. We took it to the closest nomadic family, thinking maybe it was some sort of good omen or something. Instead, they looked at us like a couple of crazy, dirty, hairy white boys with a damn bird. They told us we could come in for tea, but we had to take the bird out of their camp. We walked back to the prearranged pickup point a few days later. Tim needed to catch a flight back home, his visa was up, and we were running low on supplies. A few days after Tim flew home, I too flew home.

14
The Beginning

What makes a person brave is not going forward without fear, what makes them brave is going forward despite the fear. My relationship with fear is one of the greatest relationships I have nurtured into existence. I have traveling, rodeo, and wild horses to thank for forging this relationship. Traveling takes a person far outside their comfort zone if done correctly. Once you have left your comfort zone, you have many options for dealing with that vast new area in your mind.

Cry, panic, beg, call Mom, run, hide. These are all tactics I have seen folks use. I have used some myself. You can also learn, grow, overcome, fight, roll with the punches, and develop into a better person. As with anything in life, the more time you spend in this fruitful land outside your comfort zone, the clearer it becomes. Legendary mountaineer Reinhold Messner once said, "By climbing mountains, we were not learning how big we were.

We were finding out how breakable, how weak and full of fear we are."

Traveling for me became the release I needed badly in my life. Every cowboy speaks of the desire to see what's over the next hill, what lies just beyond what the eye can see. This drove me farther and farther into wild places. If people only knew how big this world is, just how many hills there are to see the other side of, they would probably pick up a trot and start moving faster.

I chose to see brief glimpses of the cowboy world, to be a jack of all trades and a master of none. Some folks spend their entire lives mastering one craft. Some folks spend their entire lives learning about their own country better than I will ever know. People do not have to travel far and wide to find joy, admiration, and wonder in their daily lives. For me, it took distance. But, there is nothing wrong with people who find that in their lives without ever leaving their home.

I have always felt great jealousy for my friends and family, who were able to truly find happiness without wandering feet. What is life other than finding true happiness? Maybe you find happiness sitting on the back porch watching your grandkids play, maybe you find happiness riding a fresh colt on a chilly morning, or maybe you find it thousands of miles from home in a bedroll on the ground in a cow camp like me.

There is no correct answer to happiness. That's up to you and you alone. They say, "A rolling stone gathers no moss." This was the cost I paid for my travels. I had a very sparse financial existence for years in exchange for the experience of a lifetime. Many times I wondered if I should have been making wishes on satellites instead of stars.

When everything has all boiled down, I am not sure what I have done. If I have learned anything, it's that you should always keep your cinch tight and pack spare underwear. I have traveled the world. I have gone to the wildest places I could find and done the craziest things that scared me the most. When someone said to me, "Wow, you really have wanderlust," I replied to them, "No, something much worse has afflicted me. I have 'Wildlust.'" I will continue to roll on, chasing wide open spaces in faraway places, because that is my nature.

About the Author

As a child, J. B. Zielke rarely spent time around animals other than the family dog. At seven years old he was lifted onto a horse for the first time and his course in life was changed forever. A passion for horses quickly evolved into cattle, rodeo, and the cowboy way of life.

After many injuries, Zielke decided to end his lackluster bull riding career. Trying to fill the void that bull riding had left in his life he found a new passion, travel.

Immersing himself in cowboy cultures on six continents, J. B. Zielke found himself.

Find J.B. Zielke on the socials:

Instagram: @the.lost.cowboy
Facebook: The Lost Cowboy
Youtube: The Lost Cowboy

Made in the USA
Columbia, SC
13 February 2023

12166262R00226